Frances Fuller Victor

All Over Oregon and Washington

Observations on the country, its scenery, soil, climate, resources, and improvements, with an outline of its early history. Also hints to immigrants and travelers concerning routes, the cost of travel, the price of lan

Frances Fuller Victor

All Over Oregon and Washington
Observations on the country, its scenery, soil, climate, resources, and improvements, with an outline of its early history. Also hints to immigrants and travelers concerning routes, the cost of travel, the price of lan

ISBN/EAN: 9783744755559

Printed in Europe, USA, Canada, Australia, Japan

Cover: Foto ©Andreas Hilbeck / pixelio.de

More available books at **www.hansebooks.com**

OREGON AND WASHINGTON.

OBSERVATIONS ON THE COUNTRY,

Its Scenery, Soil, Climate, Resources, and Improvements,

WITH AN OUTLINE OF ITS EARLY HISTORY, AND REMARKS ON ITS GEOLOGY,
BOTANY, MINERALOGY, ETC.   ALSO, HINTS TO IMMIGRANTS AND
TRAVELERS CONCERNING ROUTES, THE COST OF
TRAVEL, THE PRICE OF LAND, ETC.

BY

MRS. FRANCES FULLER VICTOR,

AUTHOR OF " RIVER OF THE WEST."

SAN FRANCISCO:
PRINTED BY JOHN H. CARMANY & CO., 409 WASHINGTON ST.
1872.

Printed by
JOHN H. CARMANY & CO,

Stereotyped at
THE CALIFORNIA TYPE FOUNDRY.

PREFACE.

AFTER a residence of five years on the Pacific Coast—three of which were spent in Oregon—a visit of several months was made to the Atlantic States, during which time I was called upon to do volumes of talk about the West Coast, especially about Oregon and Washington,—questions concerning which every body I met was sure to ask. The great dearth of information concerning these countries there, suggested to me the need of books which should faithfully and familiarly treat of them, their natural features and resources, together with their business and social condition.

Although every summer of my residence in Oregon had heretofore been spent in excursions to different parts of the country, I resolved on my return to repeat some of my previous journeys, and make others in new directions, until all was perfectly familiar, and thoroughly understood, which related to the geography, topography, scenery, soil, climate, productions, and improvements of the several sections of the two divisions of the Northwest Coast treated of in this volume.

While it was a pleasure to me to familiarize myself with the country, the object of it was to enable me to answer, in print, all the various questions which had been asked me concerning it by Eastern people. If the reader follows my summer wanderings as here given, he will, especially with any thing of a map before him, be able to obtain quite a complete picture of all that magnificent territory, now being rapidly brought into communication with the East, through the enterprise of the several great railroad companies. Besides the general information thus given, I have thought it best to furnish many details of the condition of the agricultural districts, and their population, for the guidance of those persons who may be looking out for farms; and of the situation and population of towns, with a view to aid immigrants in the selection of homes. It is difficult to write with absolute correctness of those countries whose rapid development outruns the printer and publisher. Since this volume was put in the hands of the compositor, numerous corrections have been made; and, between that time and this, new town-sites have been laid out, and other improvements commenced, which do not appear in these pages. But these slight omissions do not affect the general faithfulness of their contents; the whole constituting an amount of information which could only be obtained, otherwise, by a considerable expenditure of time and money.

The beautiful and favored region of the North-west Coast is about to assume a commercial importance which is sure to stimulate inquiry concerning the matters herein treated of. I trust enough is contained between the covers of this book to induce the very curious to come and see for themselves.

Portland, March, 1872.

CONTENTS.

CONTENTS.

INTRODUCTION.

The River is the Soul of the land to which it belongs. Fringing its banks, floating upon its waters, are the interests, the history, and the romance of the people. Our ideas of every nation are intimately associated with our ideas of its rivers. To mention the name of one, is to suggest the characteristics of the other.

How the word Euphrates recalls the earliest ages of man's history on this globe! The Nile reminds us of a civilization on which the whole of Europe depended for whatever was enlightened or refined anterior to the Christian Era. The Tiber is rich in historic associations of the proudest empire the world ever knew. What romances of Moorish power and splendor are conjured up by the mention of the Guadalquivir! The Rhine is so enwreathed with flowers of song, that the actual history of its battlemented towers is lost from view; and yet the mention of its name gives us a satisfying conception of the ideal Germany, past and present.

So the Thames, the Rhone, the Danube, are so many words for the English, the French, and the Austrian peoples. In our own country, what different ideas attach to Connecticut, Hudson, Savannah, and Mississippi! How quickly the pictures are shifted in the stereoscope of imagination by changing Orinoco for San Joaquin, Amazon for Sacramento, or Rio de la Plata for Columbia, upon our tongues. It is not that one is longer or shorter, or wider or deeper, than another: it is that each conveys a thought of the country, the people, the history, and the commerce of its own peculiar region.

In comparison with other rivers of equal size and geographical importance, the Columbia is very little known. That generation is yet in its prime which was taught that the whole of the North-west Territory was Oregon—that it had one river, the Columbia, and one town, Portland, *situated on the Columbia.* It is the object of this volume to correct these effete notions of one of the most genial and beautiful portions of our Republic; and to connect with the name of the Columbia some proper ideas of its history, geography, commerce, and scenery, as well as to describe the extensive country which it drains.

CHAPTER I.

DISCOVERY.

FROM the year 1513, when Balboa discovered the Pacific Ocean at Panama, the navigators of Spain, and of every rival naval power which arose for the following two hundred and seventy-nine years, were searching for some strait, or river, which should furnish water communication between the two great oceans which border the American continent. The Strait of Magellan, discovered soon after the Pacific, afforded a way by which vessels could enter this ocean from the western side of the Atlantic; but it was far to the south, crooked, and dangerous. After the discovery by the English buccaneer, Drake, of the passage around Cape Horn, the search was continued with redoubled interest. Not only the Spanish and Portuguese entered into it; but the English, who had found the great inland sea of Hudson's Bay penetrating the continent toward the west, endeavored, by offering prizes, to stimulate the zeal of navigators in looking for the North-west Passage.

A rumor continued to circulate through the world, vague, mystical, and romantic, of half discoveries by one and another power; and tales, wilder than any thing but pure fiction, were soberly listened to by crowned heads—all of which went to confirm the belief in the hoped-for straits, which one pretender to discovery even went so far as to name, and give latitude and longitude. The Straits of Anian he called them; and so, all the world was looking for *Fretum Anian.*

All this agitation could not go for nothing. By dint of sailing up and down the west coast of the continent, some actual discoveries of importance were made, and other hints of things not yet discovered were received. There even appeared upon the Spanish charts the name of a river somewhere between the 40th and 50th parallels—the San Roque—supposed to be a large river, possibly the long-sought channel of communication with the Atlantic; but no account of having entered it was ever given. Then, vague mention began to be made of the " River of the West," whose latitude and longitude nobody knew.

Just before the War of the Revolution, a Colonial Captain, one Jonathan Carver, being inspired with a desire to know more of the interior of the continent, traveled as far west as the headwaters of the Mississippi. While on this tour, he heard, from the Indians with whom he conversed, some mention of other Indians to the west, who told tales of a range of mountains called Stony Mountains, and of a great river rising in them, and flowing westward to the sea, which they called *Oregon.*

After the War of the Revolution, Great Britain resumed her voyages of discovery. A fleet was fitted out to survey the North-west coast of America, which it was thought might be claimed by her on account of the voyage to it by Captain Cook some years previous. The surveys conducted by Captain Vancouver were elaborate and scientific. He, too, like those who had gone before him, was looking for the "River of the West," or the North-west Passage.

But that obtuseness of perception, which sometimes overtakes the most sharp-sighted, overtook Captain Vancouver when his vessel passed the legendary river; for it was broad daylight and clear weather, so that he saw the headlands, and still he declared that there was no river there—only a sort of a bay.

Fortunately, a sharper eye than his had scanned the same opening not long before: the eye of one of that proverbially sharp nation, the Yankee. Captain Robert Gray, sailing a vessel in the employ of a firm of Boston traders, in taking a look at the inlet, and noticing the color of the water, *did* think there was a river there, and so told the English Captain when his vessel was spoken. Finding that his impressions were treated with superior skepticism, the Yankee Captain turned back to take another look. This second observation was conclusive. He sailed in on the 11th of May, 1792.

From the log-book of the *Columbia*, Captain Gray's ship, we take the following extracts: At four o'clock, on the morning of the 11th, "beheld our desired port, bearing east-south-east, distant six leagues. At eight A.M., being a little to the windward of the entrance of the harbor, bore away, and ran in east-north-east, between the breakers, having from five to seven fathoms of water. When we were over the bar, we found this to be a large river of fresh water, up which we steered. Many canoes came alongside. At one P.M., came to, with the small bower, in ten fathoms; black and white sand. The entrance between the bars bore west-south-west, distant ten miles; the north side of the river, distant a half mile from the ship; the south side of the same, two and a half miles distant; a village on the north side of the river, west by north, distant three-quarters of a mile. Vast numbers of the natives came alongside: people employed pumping the salt water out of our water casks, in order to fill with fresh, while the ship floated in. So ends."

No, not so ends, O modest Captain Gray, of the ship *Columbia!* The end is not yet, nor will be, until all the vast territory, rich with every production of the earth, which is drained by the waters of the new-found river, shall have yielded up its illimitable wealth to distant generations.

The *Columbia's* log-book certainly does not betray any great elation of mind in her officers on reaching the "desired port." Everything is recorded calmly and simply — quite in the way of business. Only from chance expressions, and the determination to make the "desired port," does it appear that Gray's heart was set on discovering the San Roque of the Spanish navigators — the "River of the West" of the rest of mankind. No explorer he, talking grandly of "minute inspections" and of "unalterable opinions!" Only an adventurous, and, withal, a prudent trader, looking out for the main chance, and, perhaps, emulous of a little glory.

No doubt his stout heart quaked a little with excitement, as he ran in for the "opening." We could pardon him if it shrank somewhat at sight of the hungry breakers; but it must have been a poor and pulseless affair of a heart that did not give a throb of exultation, as his good ship, dashing the foam from her prow, sailed between the white lines of surf safely —through the proper channel, thank God!—out upon the broad bosom of the most magnificent of rivers.

We trust the morning was fine, and that Captain Gray had a perfect view of the noble scenery surrounding him: of a golden sunrise from a horizon fretted by the peaks of lofty hills, bearing thick unbroken forests of giant trees; of low shores embowered in flowering shrubbery; of numerous mountain-spurs putting out into the wide bay, extending miles east and west, and north and south, forming numerous other bays and coves, where boats might lie in safety from any storm outside; of other streams dividing the mountains into ridges, and pouring their tributary waters into the great river, through narrow gaps that half revealed and half concealed the fertile valleys nestled away from inquisitive eyes: and that, as he tried in vain to look beyond the dark ridge of Tongue Point, around whose foot flowed the broad, deep current whose origin was still a mystery, he realized by a prophetic sense the importance of that morning's transaction. No other reward had he in his lifetime, and we trust he had that.

From the ship's log-book, we learn that he did not leave the river for ten days, during which time the men were employed calking the pinnace; paying the ship's side with tar, painting the same, and doing such carpenter-work as was needed to put the vessel in repair after her long voyage out from Boston. All this time, "vast numbers" of natives were alongside continually, and the Captain must have driven a thriving trade in furs, salmon, and the like. On the 14th, he sailed up the river about fifteen miles, getting aground just above Tongue Point, where he mistook the channel among the many islands; but the ship "coming off without any assistance," he dropped down to a better anchoring place.

On the 15th, in the afternoon, Captain Gray, and Mr. Hoskins, the first officer, "went on shore in the jolly-boat, to take a short view of the country." On the 16th, the ship returned to her first position off the Chinook village, and was again surrounded by the canoes of that people. Just as it was seventy-nine years ago, the Chinook village remains to-day—a cluster of huts on the north side of the river; but its people are no longer numerous. It is rare to see a single canoe, where they used to swarm in fleets on this portion of the river.

Captain Gray was thinking of getting to sea again by the 18th; but on standing down the river toward the bar, the wind came light and fluttering, and again the anchor was dropped. He must now decide upon a name for this great stream, which from its volume he knew must come from the heart of the continent. The log of the 19th says: "Fresh and clear weather. Early a number of canoes came alongside: seamen and tradesmen employed in their various departments. Captain Gray gave the river the name of Columbia's River; and the north side of the entrance, Cape Hancock; that on the south side, Point Adams."

On the 20th of May, the ship took up anchor, made sail, and stood down the river, coming, as the following extract will show, near being wrecked: "At two, the wind left us, we being on the bar with a very strong tide, which set on the breakers. It was now not possible to get out without a breeze to shoot her across the tide; so, we were obliged to bring up in three and a half fathoms, the tide running five knots.

At three-quarters past two, a fresh wind came in from seaward; we immediately came to sail and beat over the bar, having from five to seven fathoms water in the channel. At five P.M., we were out, clear of all the bars, and in twenty fathoms water."

Captain Gray proceeded from Columbia's River to Nootka Sound, a favorite harbor for trading vessels, but in dispute at that time between Spain and Great Britain. Here he reported his discovery to the Spanish Comandante, Quadra, and gave him a copy of his charts. In the controversy which afterward happened between Great Britain and the United States, concerning the title to the Oregon territory, the value of this precaution became apparent: for in that controversy the Comandante's evidence destroyed the pretensions of Vancouver's lieutenant, Broughton, who, on having heard of Gray's discovery, returned to the Columbia River, and made a survey of it up as far as the mouth of the Wallamet, founding upon this survey the claim of Great Britain to a discovery-title. The subterfuge was resorted to of denying that the Columbia was a *river* below Tongue Point; but it was claimed that it was an inlet or sound. Were it not a fact patent to every one, that a river must extend as far as the force of its current is felt, the pretense would still be perfectly transparent, since Gray must have passed Tongue Point, and been in what Broughton claimed to be the actual river before he grounded. Years afterward, the log-book of the obscure Yankee trader, and the evidence of Comandante Quadra, overbore all strained pretenses, and manifest destiny made Oregon and its great river a portion of the American Republic.

Captain Robert Gray was the first man to carry the flag of the United States around the world, having, in the spring of 1792, just returned from a voyage from Nootka to Canton, and from Canton to Boston, by way of the Cape of Good Hope. He continued to command a trading vessel up to the time of his death, in 1809. Gray's Harbor, on the coast of Washington Territory, was discovered and named by him, the name remaining as a memorial. Ought he not have some other?

In October, 1792, Vancouver having finished the survey of Puget Sound, in which the Spanish fleet was also engaged, Broughton was dispatched to the Columbia River, with the *Chatham*, which grounded just inside Cape Hancock; was got off, and anchored in a small bay on the north side of the river, known as Baker's Bay. In this cove he found, to his surprise, another vessel, the brig *Jenny*, from Bristol, England, commanded by Captain Baker, from whom he had parted in Nootka Sound. The cove was thence named Baker's Bay. From this time, the Columbia continued to be visited by trading vessels up to the war of 1812, which interrupted this sort of traffic for the time.

CHAPTER II.

In the commencement of the present century, when we paid for our teas and silks with seal-skins, cocoanut oil, and sandal-wood, not to mention turtle and abalone shells, the United States were bounded by the British provinces on the north, by the Spanish possessions, called Florida, on the south, and by the French possessions, called Louisiana, on the west. Our sea-coast extended only from the northern boundary of Maine to the southern boundary of Georgia; and the Mississippi River represented our western water-front, although the settlements in that part of our territory were chiefly French. Beyond the Mississippi was an expanse of country whose extent was undreamed of, as its geographical configuration was unknown. The explorations of the British Fur Companies in the north had revealed the existence of high mountains, and great rivers in that direction; while the little knowledge obtained of the sources of the Missouri, the Columbia, and the Colorado, together with the immense volumes of these rivers, at so great an apparent distance from their springs, was sufficient to stimulate public inquiry and scientific research. How long such inquiry would have been deferred, but for a fortunate turn in the public affairs of the United States, can only be conjectured.

Our young Republic had barely established her independence, and shaken off the lingering grasp of

Great Britain from the forts and towns bordering on the Great Lakes ; had only just begun to feel the young giant's blood in her veins, and to trust her own strength when measured with that of an older and adroit foe—when the nineteenth century dawned, in which so much has already been accomplished, though its seventh decade is but just completed.

The first event of importance marking this period, and bearing upon the history of Oregon, was the purchase from France of the *Louisiana* territory. This was a vast area of country, drained by the waters of the Mississippi, and originally settled by the French from Canada, especially in its more northern parts. Notwithstanding the Spaniards had discovered the lower Mississippi, and claimed a great extent of country under the general name of Florida, King Louis XIV. of France, in consideration of the fact that the region of the Mississippi remained unoccupied by Spain, while it was gradually being settled by his own people, thought proper to grant to Antoine Crozat, in 1712, the exclusive trade of the whole of southern Louisiana, the country included in this grant extending "from the sea-shore to the Illinois, together with the Rivers St. Philip (the Missouri), and the St. Jerome (the Ohio), with all the countries, territories, lakes in the land, and rivers emptying directly or indirectly into that part of the River St. Louis" (the Mississippi). Spain not being able to offer any successful opposition to this extensive land-grant of territories to which she laid claim by the right of discovery, Crozat remained in possession of Louisiana, under the general government of New France, until 1717, when, not finding the principality such a mine of wealth as he expected it to be, and having suffered a great private grief which took away the love of power, he relinquished his title, and Louisiana reverted to the crown. The Illinois country was afterward added to the original Louisiana territory, and the whole once more granted to *Law's Mississippi Company*, which company held it until 1732, when, the bubble of speculation being hopelessly flattened, Louisiana once more reverted to the French crown, and remained a French province until 1769.

In the meantime, however, certain negotiations were being carried forward which were to decide the future boundaries of the United States. In 1762, on the 3d of November, a convention was held at Paris, to settle the preliminaries of peace between France and Spain on the one part, and England and Portugal on the other, in which convention it was agreed that France should cede to Spain "all the country known under the name of Louisiana, as also New Orleans and the island on which that city is situated." On the 23d of the same month, this cession was formally concluded, giving to Spain, with the consent of Great Britain and Portugal, all the country drained by the Mississippi and its tributaries, except a small portion north of the Illinois country, which was never mentioned in the boundaries of Louisiana.

In less than three months after the cession of Louisiana to Spain, a treaty was concluded in Paris between the same high contracting parties, by which Great Britain obtained from France Canada, and from Spain Florida, and that portion of Louisiana east of a line drawn along the middle of the Mississippi, "from its source to the River Iberville, and thence along the middle of the Iberville, and the Lakes Maurepas and Pontchartrain, to the sea."

This treaty defined the limits of the territories belonging to Great. Britain, and set aside any former grants of English Kings, made when the extent of the continent was not even surmised. Thus, at the close of the Revolutionary War, when the United States became heirs of all the British possessions south of Canada, their western boundary, as before mentioned, was the Mississippi, as far south as the River Iberville and Lake Pontchartrain—New Orleans and the mouths of the Mississippi belonging to Spain.

Florida, during the time it was in the hands of Great Britain, had been divided into two provinces, separated by the Appalachicola River, and settled chiefly by emigrants from the south of Europe, to whose numbers, also, a few Carolinians were added. This colony of foreigners was used, in connection with the savage natives of Florida, with great effect against the southern colonies during the War of Independence. However, while they were directing their energies against Georgia, the Spaniards of Louisiana seized the opportunity for making incursions into these nondescript British provinces, and captured their chief towns, thereby rendering them worthless to Great Britain; and in 1783, Florida was retroceded to Spain, in whose hands it was in the beginning of the nineteenth century, then forming the southern boundary of the United States

In all these transactions the limits of neither Florida nor Louisiana had ever been distinctly defined; the southern boundaries of the latter infringing upon the western boundaries of the former territory. In 1800, when Spain retroceded Louisiana to France, it was described in the treaty as being the "same in extent that it now is in the hands of Spain, *and that it had been when France possessed it*"—that is, embracing the whole territory drained by the Mississippi and its tributaries, "directly or indirectly."

In 1803, April 30th, this vast extent of country was ceded to the United States by France, "with all its rights and appurtenances, as fully, and in the same manner, as they had been acquired by the French Republic," by the retrocession of Spain. By this transfer on the part of France, the Spanish government seemed at first disposed to be offended, and to offer obstacles to the settlement of the Americans in their newly acquired territory. Doubtless, this feeling arose from the unsettled condition of the boundary questions, and a fear that the United States would, as they did, demand the surrender of the whole of the original territory of Louisiana, called for by the treaty. Spain then undertook to prove that the pretensions of France to any territories west of the Mississippi could not be supported, and that the French settlements were only tolerated by Spain for the sake of peace. Such a discrepancy between the views of the two nations forbade negotiation at that time, and the matter rested, not to be revived until 1817. In the meantime, however, the United States, in 1811, feeling the necessity of holding the principal posts in the disputed territory against all other powers, took possession of the country west of the Perdido River, which was understood to be the western limit of Florida. But a British expedition having fitted out from Pensacola during the second war with Great Britain, the United States sent General Jackson to capture it, which he did in 1814, and again in 1818, as also the Fort of St. Mark. These repeated demonstrations of the spirit of the United States led to further and more successful negotiations with Spain, which power finally ceded to the American Government the whole of the territory claimed to belong to Florida, February 22, 1819, the boundaries being settled as follows :

"ARTICLE 3. The boundary line between the two countries west of the Mississippi shall begin on the Gulf of Mexico, at the mouth of the River Sabine, in the sea, continuing north, along the western bank of that river, to the 23d degree of latitude; thence, by a line due north, to the degree of latitude where it strikes the Rio Roxo of Natchitoches, or Red River; then, following the course of the Rio Roxo westward, to the degree of longitude 100 west from London and 23 from Washington; then, crossing said Red River, and running thence, by a line due north, to the River Arkansas ; thence, following the course of the southern bank of the Arkansas, to its source in latitude 42 north ; and thence, by that parallel of latitude, to the South Sea."   .   .

Other particulars are added in the article quoted, the meaning of which is the same as the foregoing: intended to fix the western boundary of the United States, as regarded the Spanish possessions, and the eastern and northern boundaries of the Spanish possessions, as regarded the United States.

Spain had never withdrawn her pretensions to the North-west Coast; but, being unable to colonize this distant territory, and still less able to hold it by garrisons in forts, she tacitly relinquished her claim to the United States, by making the forty-second parallel the northern limit of her possessions on the Pacific. The United States were then at liberty to take possession of that which Spain relinquished in their favor; in fact, had the same right to this remote territory that they had to the Florida and Louisiana territories, which were obtained by treaty from nations claiming them by the right of discovery.

But the claims of the United States to the so-called Oregon territory had even better foundations than this, if it be considered that Spain had actually abandoned her possessions in the north-west; for, in that case, the Oregon territory was theirs by the right of discovery and actual occupation, as well as by contiguity, by treaty, etc. At the time that Gray discovered and named Columbia's River, important as the discovery was, it awakened but little thought in the American mind; because, as yet, we had not acquired Louisiana, stretching to the Rocky Mountains, nor even secured the coast of the Gulf of Mexico, which was much more of an object, at that time, than the coast of the Pacific. However, when Louisiana became ours, the national mind awoke to the splendid possibilities of the nation's future. It was not for naught that a company of Boston merchants had opened a trade between China and the North-west Coast; albeit, their captains gathered up trinkets of all sorts to add to their stock in trade, should furs fall short of the market. Not in vain had the prying Boston traders peered into all inlets, bays, and rivers on the North-west Coast. When it came to discovery-rights, they had more claims than any people, the original discoverers excepted; and when Captain Vancouver's journal was published, it only convinced them that they should be fools not to profit by what it was so evidently fair they should profit by, though they did not quite see the way clear to the occupancy of the country which Columbia's River was believed to drain, nor of the islands and bays which their trading ships had explored. If Spain chose to hold possession of these coasts, they would not interfere; but if Great Britain attempted to override both Spain and America, in laying claim to the Pacific side of the continent, something might be done by way of preventing this attempt.

Such must have been the thought, half-indulged, half-repressed, in the American mind, previous to the acquisition of the great Louisiana territory. After that acquisition, it became more decided. The fact that Gray had discovered the great River of the West, which for a century had been sought after, the increasing evidences of the incapacity of Spain to hold this far-off coast against intruders, the feeling that Great Britain had no right to the countries she had so pompously taken possession of in the face of their actual discoverers—all these reasons, joined to the probable fact that the Louisiana territory bordered upon that drained by the great western river, which an American was first to enter and explore, at length shaped the policy of a few leading minds among American statesmen.

It was even contended by some, that, as the western boundary of Louisiana had never been fixed, and, indeed, was entirely unknown—since the Missouri and its tributaries had never been explored—the limits of the newly-acquired territory might be considered as extending to the Pacific; and if one were to consult the old French maps for confirmation of such an opinion, he would find *New France*, to which Louisiana belonged, extending from ocean to ocean. Yet, a perfectly candid mind would ignore the authority of maps drawn from rumor and imagination, and wish to found an opinion upon facts. It was to secure such facts and to carry out, as far as possible, the lately formed policy of leading statesmen, that President Jefferson, even before the transfer of Louisiana was completed, addressed a confidential message to Congress, urging that means should be immediately taken to explore the sources of the Missouri and the Platte, and to ascertain whether the Columbia, the Oregon, the Colorado, or any other river, offered a direct and practicable water-communication across the continent, for purposes of commerce. The suggestions of the President being approved, commissions were issued to Captains Merriwether Lewis and William Clarke to perform this service. Captain Lewis made immediate preparations, and, by the time that the news of the ratification of the treaty had been received, was ready to commence his journey to the unknown West.

It was already summer when this news was received, and, although the party were ready to advance into the Indian country, it was too late to accomplish much of their journey before winter; besides which, some delay occurring in the surrender of the country west of the Mississippi, the party were not able to cross that river until December—in consequence of which detention, the ascent of the Missouri could not be undertaken before the middle of May, of the following year. The exploring party consisted of but forty-four men—an insignificant force to send into an Indian country—yet, perhaps, all the safer for its insignificance. They had to make the ascent against the current of the Mad River in boats, three of which sufficed to accommodate this adventurous expedition. By the end of October, they had arrived in the Mandan country, near the forty-eighth degree of latitude, or sixteen hundred miles from the Mississippi, where they made their winter camp. As every school-library is furnished with the printed journal of Lewis and Clarke, it is unnecessary to dwell upon the incidents of their memorable journey across the continent. It is only with its results that we have to deal in this sketch.

One of its results was developed at this early period, or during their stay at the Mandan village: which was, to alarm the North-west Fur Company, and, through them, the English Government, as to the designs of the Americans concerning the northern coast of the Pacific. It has been before stated, that the North-west Company had been compelled reluctantly to resign the posts along the Great Lakes, belonging to the United States, after the Revolutionary War. They still continued to hunt and trap, and had established their trading-posts in all that country lying about the head-waters of the Mississippi; and their employees were scattered throughout the region east of the Missouri, and west of the Lakes—even having penetrated, on one occasion, to the foot of the Rocky Mountains.

It happened that, while Lewis and Clarke were at the Mandan villages, the fact of their visit, and the object of it, which had been explained to the Indians, were communicated to some members of the North-west Company, who had a post about three days' journey from there. So much alarmed was Mr. Chaboillez, who resided at this post, that he wrote immediately to another partner, Mr. D. W. Harmon, a native of New England; and, upon receiving a visit from him, urged Mr. Harmon to set out in the following spring upon the same route pursued by Lewis and Clarke, accompanied by Indian guides, doubtless with the intention of arriving at the head-waters of the Missouri, in advance of the American expedition; but, in this praiseworthy strife for precedence they were in this instance defeated—Mr. Harmon proceeding no further than the Mandan villages, while Lewis and Clarke prosecuted their undertaking with diligence, leaving the Mandan country on the 7th of April, 1805, and arriving at the Great Falls of the Missouri on the 13th of June. The reader need not be reminded of the difficulties attending such a journey as the one undertaken by our exploring party: when, the course of navigation being interrupted, boats had to be abandoned, toilsome portages made, new boats constructed, and all the novel hardships of the wilderness endured. Such tests of courage have been encountered by thousands since that time, in the settlement of the Pacific Coast; but that fact does not lessen the glory which attaches to the fame of the great pioneers commissioned to discover the hidden sources of America's greatest rivers. Those faithful services secured to us inestimable blessings, in extended territories, salubrious climates, and exhaustless wealth of natural resources.

Lewis and Clarke, having re-embarked in canoes hollowed out of logs, arrived at the *Gate of the Mountains* on the 19th of July, in the very neighborhood where thousands of men are to-day probing the earth for her concealed treasures of gold and silver. Proceeding on to the several forks of the Missouri—the Jefferson, the Madison, and the Gallatin—and finding themselves in the midst of the mountains, the two captains left a portion of their men to explore the largest of these, while they, with the remainder of the party, pushed on through the mountains until they came to streams flowing toward the west. At this intimation that their labors were about to be crowned with success, they rejoined their party at the head of the Jefferson Fork, and prepared for the rugged work of crossing that majestic range, now become so familiar. Concealing their goods and canoes in *caches*, after the fashion of all knowing mountaineers, and being furnished with horses and guides by the Shoshones, or Snake Indians, whose later hostility to the whites makes us wonder at their early friendship for Lewis and Clarke, the party commenced the passage of the Rocky Mountains on the 30th of August. Severe was their toil, and great were the sufferings they endured from hunger and cold; but, at length, their trials passed, they arrived at a stream on which their Indian guides allowed them to embark. This was the Clearwater River, the banks of which have since become historic ground.

The party were glad again to be able to resume water navigation, and hastened to build their canoes, and place their horses in charge of the Chopunish, or Nez Perce tribe of Indians, whose extraordinary fidelity to the treaty formed at that time with Lewis and Clarke is one of the wonders of history. On the 7th of October, they began to descend the Clearwater, and three days later entered upon that great branch of the Columbia, whose springs they had, indeed, tasted in the mountains, but upon whose bosom no party of civilized men had ever before embarked.

Men are apt to dwell with enthusiasm upon the pride of a conqueror; but, certainly, there must be that in the exultation of a discoverer, which is far more pure, elevated, and happifying. To have succeeded, by patient research and energetic toil, in securing that which others secure by blood and devastation only, is justly a subject of self-congratulation, as it is also deserving of praise. The choicest wine, from the costliest chalice, could hardly have been so sweet to the taste of our hardy exploring party, as the ice-cold draught of living water dipped from the mountain reservoirs whose streams "flowed toward the west." With equal pride must they have launched their frail canoes on that river which now bears the name of the chief of the expedition. As they descended to the junction with the northern branch, and found themselves at last fairly embarked on the main Columbia, when they beheld the beauty and magnitude of this King of Rivers, and remembered that their errand, so successfully carried out, was to find a "highway for commerce," their toils and privations must have appeared to them rather in the light of pleasures than of griefs. As the first party of white men to pass through the magnificent mountain-gap where the great river breaks through the Cascade Range, and to meet the tides of the Pacific just on the westward side, the party of Lewis and Clarke have won, and ever must retain, an honorable renown.

The voyage from this point to the mouth of the Columbia was soon accomplished. On the 15th of November, the expedition landed at Cape Hancock, commonly called "Disappointment," on the north side of the river, having traveled a distance of more than four thousand miles from the Mississippi River. The rainy season, which usually sets in about the 18th of November, had already commenced, so that our explorers had some difficulty in finding a suitable winter camping-ground. At first, they tried the peninsula north of Cape Hancock, but were driven from their ground by the floods. Then they resorted to the south side of the river, somewhat farther back from the ocean, building a log fort on a small stream which is still called "Lewis and Clarke River." There they ·

contrived to pass the winter without actual starvation, though they were often threatened with it, from the difficulty of obtaining food at this season of the year. Game was scarce, except in the coast mountains, which are very rugged and thickly wooded; while fishing could not be carried on successfully except with other boats than their slight canoes, which were entirely unfit for the winter winds and waves of the lower Columbia. The Indians, among whom they wintered, called themselves "Clatsops," and were sufficiently friendly, but had no food to spare save at the very highest prices. The Chinooks on the north side of the Columbia, the same people Captain Gray had traded with thirteen years before, were equally exorbitant in their prices, and exercised a monopoly of the necessaries of life quite equal to that of the most practiced extortionists.

Nothing could be effected in the way of explorations of the country during the winter of 1805–6, on account of the rains, which were constant and excessive; and the party, however unwillingly, remained at Fort Clatsop until the middle of March, going no farther away than to Cape Lookout, about fifty miles down the coast. As soon as the rainy season had closed, Lewis and Clarke re-embarked their men, and returned up the river, surveying the shores in their voyage. In this passage they discovered the Cowlitz River, the principal tributary emptying into the Columbia from the north side anywhere west of the Cascades. The Wallamet River was also discovered, but remained unexplored, from the anxiety of the expedition to return to the United States.

By the middle of April, the party had abandoned their canoes at the gap in the Cascade Mountains, where the river forms dangerous rapids; and, purchasing Indian horses, continued the journey on horseback to the Nez Perces country, where these faithful allies met them on their return, not with friendship only, but with the animals confided to their care the preceding autumn—an example of Indian integrity worthy of mention, and, as it proved, indicative of a character shown in the events of succeeding years.

After crossing the Rocky Mountains to Clarke's River, the two leaders of the expedition separated—Captain Lewis going northward, down the Clarke River, and Captain Clarke proceeding toward its source. On the 12th of August, the two captains met at the mouth of the Yellowstone, having explored that river, as well as the Clarke, and traversed a great extent of country then unknown to white men, but where white men, to-day, are suffering the flushes and the rigors of that most infectious and fatal complaint—the gold-fever—in the Territory of Montana.

At about the mouth of the Maria River, Captain Lewis had an encounter with the Blackfeet, the most savage and dreaded of the mountain tribes. In this conflict one of the Indians was killed, which caused the others to desist at that time; yet, no doubt, many a white man's scalp has been taken in revenge, according to savage custom—and the wonder still remains that the party escaped alive out of the country.

After re-uniting their forces—their mission being accomplished—the expedition once more embarked on the Missouri River, and arrived at St. Louis September 23d, having traveled in less than three years, by canoe and saddle, carrying their own supplies, more than nine thousand miles.

Of the results of the expedition of Lewis and Clarke, it may be said, that it was the first great act, wisely conceived and well executed, which secured the Oregon territory to the United States. It was the beginning, too, of a struggle for possession between this country and Great Britain, dating from the meeting of the North-west Company's men with the men of the American expedition at the Mandan villages. Happily, all these struggles for precedence are matters of past history now; and, to-day, both English and American citizens seek and find homes on Oregon soil, where, according to a wise act of Congress, one may be had for the taking.

The first attempt that was made to form a settlement on the Columbia River was by the Winship brothers, in 1810. On the 7th of July, 1809, there sailed from Boston two ships—the *O'Cain*, Captain Jonathan Winship, and the *Albatross*, Captain Nathan Winship. The *O'Cain* proceeded direct to California, to trade out a cargo of goods with the *padres* of the Missions, and their converts; and the *Albatross* sailed for the Sandwich Islands, with twenty-five persons on board. At the Islands she provisioned, and took on board twenty-five more men, leaving port for the Columbia, March 25th, 1810.

Arriving in the river early in the spring, Captain Winship cruised along up, for ten days, finally selecting a site on the south side, about forty miles from its mouth, and opposite the place now known as "Oak Point," though its name is borrowed from Captain Winship's place. Here he commenced founding an establishment, and, for a time, every thing progressed satisfactorily. A tract of ground, being cleared, was planted with vegetables; a building was erected; and, while the river banks were gay with the blossoming shrubbery of early summer, our captain and his fifty workmen rejoiced in the promise of a speedy consummation of their plans of colonization. Their hopes, however, were soon overthrown by an unlooked-for occurrence; and the daring pioneers, who feared the face of neither man nor beast in all that wilderness, found themselves confronted with an adversary against which it was useless to contend. The snows had melted in the mountains a thousand miles eastward, and the summer flood came down upon their new plantation, washing the seeds out of the earth and covering the floors of their houses two feet deep with water— demonstrating conclusively the unfitness of the site selected for their settlement.

Without doubt, this company of adventurers were by turns wroth and sorrowful. Their seeds were lost; their residences made uninhabitable, even had they desired to remain, which they did not. Captain Winship at once re-embarked his men, and sailed for California to consult with his brother. Here he was met by the intelligence of the formation of the Pacific Fur Company, with John Jacob Astor at its head, and the intention of this company to occupy the Columbia River. Competition with so powerful an association was not to be thought of, and the brothers Winship abandoned their enterprise. As men of large ideas and fearless action, they should be remembered in connection with the history of the Columbia River.

In March of the following year, that portion of Mr. Astor's expedition which was to come by sea, did arrive on the Columbia—not, however, without the loss of eight men on the bar, through the impatience and overbearing temper of the commander of the *Tonquin*, Captain Thorne. Subsequently, the Indians of the Straits of Fuca destroyed the *Tonquin*, massacring all her officers and crew, twenty-three in number. The land expedition suffered incredible hardships: supply vessels failed to arrive; war with Great Britain broke out, preventing Mr. Astor from carrying out his plans; the Canadian partners took advantage of the situation to betray Mr. Astor's interests; and, after two years of hope deferred, the establishment at Astoria was sold out to a British company, and the enterprise abandoned—the place having been "captured" by the British.

After the close of the war of 1812, Astoria was restored to the United States, and Mr. Astor would have renewed his enterprise, notwithstanding his heavy losses, had Congress guaranteed him protection, and lent its aid; but the Government pursued a cautious policy at this time, and the Oregon territory remained in the hands of the British fur traders exclusively for the twenty years following, notwithstanding a treaty of joint occupation.

To follow the chain of events, and record the incidents of a long struggle between Great Britain and the United States to substantiate a claim to Oregon, is the work of the historian. Enough for us, that we know which claim prevailed; and we shall proceed to the more congenial contemplation of the physical features which the country presents, touching lightly now and then upon its history, as tourists may.

CHAPTER III.

WHERE the Columbia meets the sea, in an almost continuous line of surf, is some distance outside the capes; but from the one to the other of these—that is, from Cape Hancock to Point Adams—is seven miles. Should the sea be calm on making the entrance, nothing more than a long, white line will indicate the bar. If the wind be fresh, the surf will dash up handsomely; and if it be stormy, great walls of foam will rear themselves threateningly on either side, and your breath will be abated while the quivering ship, with a most "uneasy motion," plunges into the thick of it, dashes through the white-crested tumult, and emerges triumphantly upon the smooth bosom of the river.

Of the two channels, the south is most used. Should you happen to go in by the north one, you will find yourself pretty close under a handsome promontory, with a white tower, in which a first-class Fresnel-light is burning from sunset to sunrise, all the year round. This promontory is the Cape Hancock of Captain Gray and the United States Government, and the Cape Disappointment of the English navigators and of common usage, since the long residence in the country of the Hudson's Bay Company.

The steamers of the North Pacific Transportation Company will not land you before reaching Astoria, a dozen miles inside the bar. But, for this once, we will "subsidize" our captain with many fair words, and persuade him to send us ashore in a ship's boat, that we may miss nothing in our voyage up this river we have come a long way to see.

As we round the base of the cape, we find ourselves in a pretty little harbor called Baker's Bay, with an island or two in it, and surrounded by heights of sloping ground covered with a dense growth of spruce, fir, and hemlock, with many varieties of lesser trees and shrubs. Along the strip of low land, crescent-shaped and edged with a sandy beach, are the officers' quarters and soldiers' barracks; for the cape has been fortified, and has three powerful batteries on the channel side. Nearest of all is the residence of the light-house keeper—a modest mansion under the shelter of the cape.

At this place we will call and get our bearings. We wish to pay our respects to the post-commander, and have the quarters pointed out to us. That formality—a very pleasant one—disposed of, we gladly accept a proffered escort to the fortifications. If the day be warm, we take the path through the thick woods, winding around and about up to the top of the promontory. What fine trees! What a dense and luxuriant undergrowth!

Sauntering, pulling ferns and wild vines, exclaiming at the shadows, the coolness, the magnificence of the forests, we come at last to the summit, and emerge into open ground. Here all is military precision and neatness: graveled walks, grassy slopes and terraces, whitened walls. As for the guns and earth-works, they are of the first order. When we have done with these, we turn eagerly to gaze at the sea; to watch the restless surf dashing itself against the bar; to catch that wonderful monotone—"ever, forever."

The fascination of looking and listening would keep us long spell-bound; but our escort, who understands the symptoms, politely compels us "to move on," and directly—very opportunely—we are confronted with the light-house keeper, who offers to show us his tower and light. Clambering up and up, at last we stand within the great lantern, with its intense reflections; and hear all about the life of its keeper—how he scours and polishes by day, and tends the burning oil by night. When we ask him if the storm-winds do not threaten his tower, he shakes his head and smiles, and says, it is an "eerie" place up there when the sou'-westers are blowing. But, somehow, he likes it; he would not like to leave his place for another.

Then we climb a little higher, going out upon the iron balcony, where the keeper stands to do his outside polishing of the glass. The view is grand; but what charms us most, is a miniature landscape reflected in one of the facets of the lantern. It is a complete copy of the north-western shore of the cape, a hundred times more perfect and beautiful than a painter could make it, with the features of a score of rods concentrated into a picture of a dozen inches in diameter, with the real life, and motion, and atmosphere of Nature in it. While you gaze enchanted, the surf creeps up the sandy beach, the sea-birds circle about the rocks, the giant firs move gently in the breeze, shadows flit over the sea, a cloud moves in the sky; in short, it is the loveliest picture your eyes ever rested on.

The friendly keeper explains to you, as you turn to look up the coast, that the beach north of the cape extends, in one unbroken level, about twenty miles; and that it is a long, narrow neck, divided from the main-land by an arm of Shoalwater Bay, extending almost down to the light-house. A splendid drive down from the bay! It is in the sandy marshes up along this arm of Shoalwater Bay, too, that we may go to find cranberries.

When we ask, "What does he do when the thick fogs hang over the coast?" he shows us a great bell, which, when the machinery is wound up, tolls, tolls, tolls, solemnly in the darkness, to warn vessels off the coast. "But," he says, "it is not large enough, and can not be heard any great distance. Vessels usually keep out to sea in a fog, and ring their own bells to keep off other vessels."

Then he shows us, at our request, Peacock Spit, where the United States vessel of that name was wrecked, in 1841; and the South Spit, nearly two miles outside the cape, where the *Shark*, another United States vessel, was lost in 1846. The bones of many a gallant sailor, and many a noble ship, are laid on the sands, not half a dozen miles from the spot where we now stand and look at a tranquil ocean. Nor was it in storms that these shipping disasters happened. It was the treacherous *calm* that met them on the bar, when the current or the tide carried them upon the sands, where they lay helpless until the flood-tide met the current, and the ship was broken up in the breakers. Pilotage and steam have done away with shipwrecks on the bar.

We are glad to think that it is so. Having exhausted local topics for conversation, we descend the winding stairs, which remind us of those in the "Spider and the Fly"—so hard are they to "come down again." How still and warm it is down under the shelter of the earth-works! Descending by the military road, which

is shorter than the one we came by, we come out near
the life-boat house, and, being invited, go in to look at
it.   It seems well furnished and commodious, and we
are told it is safe, but, happily, has seldom been needed.
Lastly, we take a look at the fishing-tackle, with which
the light-house keeper goes out to troll for salmon.
Glorious sport!   The great, delicious fellows, to be
caught by a fly!   But we, humans, need not sermonize
about being taken by small bait!

Baker's Bay is not without its little history; albeit,
it is nothing romantic.   In 1850, a company conceived
the plan of building up a city, under shelter of the cape,
and expended a hundred thousand dollars, more or less,
before they became aware of the fruitlessness of their
undertaking.   By mistake, portions of their improvements were placed on the Government Reserve, to
which, of course, they could have no title.   Yet, this
error, although a hinderance, was not the real cause of
the company's failure, which was founded in the ineligibility of the situation for a town of importance.
Nothing remains of the buildings there erected, their
sites being already grown over with a young forest of
alders, spruce, and hemlock.

There being nothing more of interest to be seen at
the cape, we take the little steamer *U. S. Grant*, which
has run over from Astoria with the mail for the garrison, for Point Adams on the opposite side of the river.   The wind has freshened, and the steamer rolls a
good deal, the river here feeling the ocean-breezes
very sensibly.   Such is its expanse, that, although our
course brings us off Chinook Point, we have but an in-
distinct view of it.   Not as it was seventy years ago—a
populous Indian village; the dwellings of white settlers
are now overshadowing the ancient wigwams.   Even its

burial-ground—its *memelose illihee*, or "land of spirits"—is profaned. Alas! nothing of one race is sacred to another; least-of all, are the poor Indians' bones sacred to white men.

Several localities are pointed out to us, while we cross the river; but, at this distance, we can not see much more than that to the north of us is a range of high, wooded bluffs, with a narrow strip of level ground along the river, more or less inhabited. That which does attract our attention is Sand Island, close to which we pass. It is scarcely above the level of the water, at mean tide, and presents a waste of sand, in which a few dead trees are embedded. It is fringed with a colony of eagles, who sit motionless, but keen-eyed, watching for their prey—their pre-emptive title being disputed only by a shoal of seals, whose antics furnish a pleasing contrast to the gravity of their feathered rivals. In little more than half an hour, we are landed at Fort Stevens, on Point Adams.

There is nothing handsome in the situation of Fort Stevens. It occupies a low, sandy plain, and is just a little inside of the actual point of this cape; but the fort itself is one of the strongest and best-armed on the Pacific Coast. Its shape is a nonagon, surrounded by a ditch, thirty feet wide. This ditch is again surrounded by earth-works, intended to protect the wall of the fort, from which rise the earth-works supporting the ordnance. Viewed from the outside, nothing is seen but the gently inclined banks of earth, smoothly sodded. The officers' quarters, outside the fort, are very pleasant; and, although there is nothing attractive in the appearance of the fort, or its surroundings, it is a pleasant enough place to those who have the good fortune to have the *entree* of its society.

The view from the embankment is extensive, commanding the entrance to the river, the opposite fortifications, and the handsome highlands of the north side, as well as a portion of Young's Bay. A system of signals is established between the two forts, and signal-practice is made a portion of the daily duty of the officers. Standing on this eminence, our curiosity is excited, to know why a certain small sailing-craft keeps anchored out near the bar, and are told that it belongs to the United States Surveying Service, and that its business is to observe the tides and currents on this station.

Point Adams is the northern projection of a sandy peninsula, formed by the Pacific Ocean and Young's Bay. It is a narrow neck of sand-ridges, or irregular sand-hills, interspersed with ponds and swamps, and thickly overgrown with spruce, hemlock, and other trees of similar species. Where the trees have been cleared away, thickets of wild roses, willows, and *spiræa* have sprung up, covering the ground.

Below this swampy point, the sand-ridges continue for sixteen miles to Tillamook Head, a promontory four or five hundred feet in height. A species of wild clover grows in the sand, flourishing until midsummer, when it is succeeded by a good crop of grass. The wild strawberry grows finely here; and, wherever cultivated, vegetables do well. This narrow sand-belt is known by the name of Clatsop Plains, and is nowhere more than a mile in width. Back of it, toward Young's Bay and Skippanon Creek, the land is heavily timbered, the timber extending back to the Coast Mountains.

Clatsop Plains, and all the level country between

them and the Coast Range, together form the county of that name. It is famous for its dairies, its strawberries, its vegetables, but, most of all, for its seabathing. No one is presumed to be in the fashion, who has not been to Clatsop Beach: therefore, to Clatsop we are going—have gone. We like the place, though it is as little like Newport or Long Branch as possible, having for an hotel a one-storied wooden building, brilliant externally with whitewash, internally not brilliant at all, nor elegantly furnished, being the residence of a family of French half-breeds. The *cuisine* is all that a Frenchman could desire; but the house and grounds are decidedly of a by-gone order of architecture and arrangement. When the house is overrun with visitors, the later comers are domiciled in tents. Perhaps it is this very lack of conventional luxury which makes the place popular; for it never is deserted during the warm season, but every year increases the number of its visitors. Sea-air, bathing, riding, hunting, good living, and the absence of those usual conventionalities which make life refined and monotonous, continue to "draw" more and more largely, so that shortly some sharp-sighted party will be found erecting the hotels and cottages of a crowded watering-place.

There are certainly here many attractions lacking in most sea-bathing resorts: a trout-stream, a forest for hunting in, where any thing may be found, from a deer to an elk, or a bear. Geese, ducks, plover, and snipe frequent the mouth of the creek, while sea-gulls, cranes, and eagles give picturesqueness to the beach-views. Three or four miles to the east, the peaks of the Coast Range fret the blue of the summer sky, a spur from which range comes down quite to the sea, in a bold promontory called Tillamook Head, closing in the southern view.

Having taken in all these features of the place, and pronounced it good, let us take the light wagon, and, driving across the plain and through the woods nearly sixteen miles, find the *Grant*—ubiquitous little steamer—waiting for us in Young's Bay. As we steam toward Astoria, the accomplished Captain of the *Grant* —the first white male child born west of the Rocky Mountains—becomes our guide, and points out the mouth of Lewis and Clarke's River, on the south side of the bay, where those hardy explorers spent the winter of 1805–6 in a log-hut, to which the severe rains confined them nearly all those dreary months, in imminent danger of starving. Not only have sixty years effaced all traces of their encampment, but a house, which stood on the same site in 1853, has quite disappeared, the site being overgrown with trees now twenty feet in height. Of a saw-mill that furnished lumber to San Francisco, in the same year, nothing now remains except immense beds of half-rotted sawdust, embedding one or two charred foundation timbers. A dense growth of vegetation covers the whole ground.

At the eastern extremity of the bay is the mouth of Young's River, a handsome stream, with densely wooded shores, and a fall, at one place, of fifty feet perpendicular, furnishing one of the attractions to boating parties of summer visitors at Astoria.

From the deck of the steamer we have a fine view of the Coast Range, and of one double peak higher than the range, which goes by the ugly misnomer of Saddle Mountain. Not snow-capped in summer, it is still very lofty and very picturesque, reminding us of "castled crags of Drachenfels." We, for our private satisfaction, name it Castle Mountain, and try to forget that it has another name.

As we round the high, wooded point which hides Astoria from sight, as it must, also, shelter it from south-west storms, we observe that the banks are covered with a most luxuriant growth of shrubs of many varieties, and promise ourselves a ramble along a just visible "trail" at an early day, in order to ascertain whether or not they are as beautiful close at hand, as they are in the distance.

Our eyes are engaged, in another moment, with some glimpses of our destined port. Very shortly, the *Grant* comes alongside a great wharf, and seeking her own slip, makes fast; and, the tide being out, we clamber up cautiously a steep incline, to the level of the Astorians.

CHAPTER IV.

THE situation of Astoria, in point of beauty, is certainly a very fine one. The neck of land occupied by the town, is made a peninsula by Young's Bay on one side and the Columbia River on the other, and points to the north-west. A small cove makes in at the east side of the neck, just back of which the ground rises much more gently and smoothly than it does a little farther toward the sea. The whole point was originally covered with heavy timber, which came quite down to high-water mark; and whatever there is unlovely in the present aspect of Astoria, arises from the roughness always attendant upon the clearing up of timbered lands.

Standing, facing the sea or the river, with your back to half-cleared lots, made unsightly by the blackened stumps of trees, the view is one of unsurpassed beauty. Toward the sea, the low, green point on which Fort Stevens stands—the Cape Frondosa (leafy cape) of the Spanish navigators—and the high one of Cape Hancock, topped by the light-house tower, mark the entrance to the river. Above them is a blue sky; between them, a blue river, celebrating eternally its union with the sea by the roar of its breakers, whose white crests are often distinctly visible. There is a sail or two in the offing, and a pilot-boat going out to bring them over the bar.

Opposite us, and distant between three and four miles, is the northern shore—a line of rounded highlands, covered with trees, with a narrow, low, and level strip of land between them and the beach. The village of Chinook is a little to the north-west; another village, Knappton, a little to the north-east. Following the opposite shore-line with the eye, as far to the east as the view extends, a considerable indentation in the shore marks Gray's Bay, where the discoverer of the river went ashore with his mate, to "view the country."

On the Astoria side the shore curves beautifully, in a north-east direction, quite to Tongue Point, four miles up the river. This point is one of the handsomest projections on the river. Connected with the main-land by a low, narrow isthmus, it rises gradually to the height of fifty or sixty feet, and is crowned with a splendid growth of trees. In the little bay formed by Tongue Point, lies the hulk of a vessel—a memento of the exciting times of 1849, when lumber was worth, in San Francisco, six hundred dollars a thousand feet.

The ship *Silvie de Grace* had come to Oregon for a cargo of the precious material, and proceeded as far as this on her return-voyage, when, through ignorance or mismanagement, she was allowed to strike on a rock, with such force that she was actually spitted, and never could be got off, even to sink. So she lies a dismantled hulk in this pretty cove, not unpicturesque, with her handsomely modeled deck half-overgrown with grass and shrubs, and the headless figure of a woman "to the fore."

Between Tongue Point and the present town is a cluster of rather dilapidated buildings, known as Up-.

per Astoria. They were erected by the first Receiver of Customs for Oregon; but the old custom-house and wharf are rapidly going to decay. Directly back of this place, begins a "military road" to the State Capital, on which was supposed to be expended, in the years 1855–6, an appropriation of $80,000. It was never fit for use, and is now quite choked up with fallen timber and a new growth of trees.

Following the curving and beautifully wooded shore back to the Astoria of to-day, we naturally inquire for the site of the Astor establishment of 1811. This is it, just back of the little bay before mentioned, where you see a long, one-storied house in a state of decay. There was built the fort of Mr. Astor's company. It consisted of a square, inclosing ninety by a hundred feet of ground, with palisades in front and rear, one of the sides protected by the warehouse fronting on a ravine, and the other by the dwelling-house and shops, with a bastion at each corner, north and south, on which were mounted four small cannon. As all the buildings were constructed of hewn logs, roofed with cedar-bark, they constituted a very good defense against the Indian arrows, especially as they were made formidable by the four small cannon.

On the 26th of September, 1811, the buildings inside the fort were completed. The dwelling-house contained a sitting-room and dining-room, with sleeping apartments for the officers and men. The warehouse and smiths' shops were also there ready for occupation. In the following year a hospital was erected; and these constituted the improvements of the Pacific Fur Company, if we except their garden, where nothing came to maturity the first year, except the radishes, turnips, and potatoes.

In the cove, in front of the fort, was built the first vessel ever launched on Oregon waters—the little schooner *Dolly*, whose frame was brought out from New York in the *Tonquin*. She proved too small for the coasting service, for which she was intended, and, like every thing else connected with this ill-starred enterprise, a failure.

In 1813 the Astoria of the Pacific Fur Company passed into the hands of the North-west Fur Company, by whom it was re-named Fort George. Afterward it passed to the Hudson's Bay Company, and was known as Fort George, until it was abandoned by them, and came once more into American possession, when it resumed its original name. Such are the changes of sixty years. Nothing now remains to remind us of these events in history, except some slight indentations in the ground where were once the cellars of the now vanished fort, and a few graves. Perhaps the only enduring memorial is the smooth turf and fine grass of civilization, which Time does not eradicate, and which grows here in strong contrast to the rank, wild grasses of the uncultivated country.

If we turn to the modern town, we find it neatly built, and containing four or five hundred inhabitants. The chief improvement going on at present, is the new custom-house—a costly, but ill-looking structure, built of sandstone from the opposite side of the river. The present custom-house is a wooden building near the river, occupying the ground chosen by the officers and men of the United States schooner *Shark*, to erect their temporary shelter upon, after the wreck on the bar, in 1846. From drift-wood and cedar planks they constructed a substantial house, which, afterwards, was turned to account by others in almost equal straits. One of its last and best uses was as a ball-room, where, on the Fourth of July, 1849, the gold-seekers on their way to California, and a company of United States artillery-men, celebrated the day with patriotic enthusiasm.

Even as late as that year, the canoes of eight hundred native warriors of the Chinooks covered the water in Astor Bay, curious, as savages always are, to watch the acts, and note the customs, of civilized men. Not a canoe is now in sight. The white race are to the red as sun to snow: as silently and surely the red men disappear, dissipated by the beams of civilization. Among those who came to gaze at the overpowering white race on that occasion, was an old Chinook chief, the number of whose years was one hundred. His picture, which some one gave to us, shows a shrewd character. So, no doubt, looked Com-com-ly, the chief whom Washington Irving describes in his "Astoria," and whose contemporary this venerable savage must have been. His sightless eyes, in his early manhood, beheld the entrance into the river of that vessel whose name it bears. Between that time and the day of his death, he saw the Columbia River tribes, which once numbered thirty thousand, decimated again and again, until they scarcely counted up one-tenth of that number.

If you ask an Astorian, what constitutes the wealth and commercial importance of his town, present and future, he will tell you, that it has a commodious harbor, with depth of water enough to accommodate vessels of the deepest draft, with good anchorage, and shelter from south-west (winter) storms. He will point to the forts at the mouth of the river, and say that they make business; to the custom-house, and

that it makes business.   He will remind you of the
pilotage of all the incoming and outgoing vessels, and
that it brings in a great deal of money.   He will point
to the villages growing up on the north side of the
river, and tell you they bring trade; that the men
employed at Knappton, in making cement, lumber,
etc., spend their wages in Astoria.

If you inquire what back country it has to support
it, he will point to Clatsop, and the valley of the Nehalem, south of it; and tell you, that it is but seventy
miles into the great valley of Western Oregon—the
Wallamet; and that a railroad is to be built into it
from Astoria, through the coast mountains.   He men-
tions, besides, that there are numerous small valleys
of streams running into the Columbia within twenty
miles, which are of the best of rich bottom-lands, and
only need opening up.   This is the Astorian's view of
his town, and we know nothing to the contrary.   Indeed, from inquiry we are convinced that there are in
the neighborhood of Astoria many elements of wealth,
both mineral and agricultural, which only require time
and capital to develop.

Having satisfied ourselves of the material prospects
of the town, let us take a friendly guide, and go upon
an exploring expedition on our own account.   We want
to go on foot around the point, by the trail through
the woods: but, no; our guide says we must not attempt it, the trail is in such a condition!   "It is low
tide, and we will go by the beach."

By the beach we go, then, stopping now and then to
fillip a jelly-fish back into the water on the end of our
alpenstock.   A beach, indeed! we had always thought
that sand, or fine gravel, at least, was essential to that
delightful thing in Nature—a beach.   But here are

bowlders, growing larger and larger as we near Young's Bay, until just at the extremity of the point they require much exertion to scramble over. But our guide is entertaining, which compensates for great exertion.

In stories of "peril by land and water," of shipwrecks, and legends of treasure-trove—that should be —he drowns all thoughts of mutiny, and we toil ahead. "To be sure there have been wrecks at the mouth of the Columbia—a century—two centuries ago." Then he takes from his pocket, where he must have placed it for this purpose, and shows to us a thin cake of bees-wax, well sanded over, which he avers was portion of the cargo of a Japanese junk, cast ashore near the Columbia in some time out of mind. When we have wondered over this, to us, singular evidence of wrecking, he produces another, in the form of a waxen tube. At this we are more stultified than before, and then are told that this was a large wax candle, such as the Japanese priest, as well as the Roman, uses to burn before altars. The wick is entirely rotted out, leaving the candle a hollow cylinder of wax.

By this self-evident explanation, we are convinced. Certain it is that for years, whenever there has been an unusually violent storm, portions of this waxen cargo are washed ashore, ground full of sand. As bees-wax is a common commodity in Japan, we see no reason to doubt that this, which the sea gives up from time to time, originally came from there. The supposition is the more natural, as the mouth of the Columbia is exactly opposite the northern extremity of that Island Empire; and a junk, once disabled, would naturally drift this way. The thing has been known to occur in later years; and that other wrecks, probably Spanish, have happened on this coast, is evidenced by the light-haired and freckle-faced natives of some portions of it farther north, discovered by the earliest traders.

Our hour of toil, at length, brings us to a pretty piece of level, grassy land away from the beach, where are lofty trees, and lower thickets of wild roses, white *spiræa*, woodbine, and mock-orange. Here, in this charming solitude, is an Indian lodge, the residence of the native Clatsop; and we have a strong desire to see its interior. Exteriorly, the Clatsop residence can not be praised for its beauty, being made of cedar planks, set upright and fastened to a square or oblong frame of poles, and roofed with cedar bark. Outside are numberless dogs, and two pretty girls, of ten and twelve years of age, with glorious great, black, smiling eyes.

Peeping inside, we see three squaws of various ages, braiding baskets and tending a baby of tender age, with two "warriors" sitting on their haunches and doing nothing; and salmon everywhere—on the fire, on the walls, overhead, dripping grease and smelling villainously, are salmon—nothing but salmon. Our guide holds a conversation with the mother of the little stranger, in jargon, which he informs us relates to the fair complexion of the *tillicum*. One of the warriors, presumed to be its papa, laughs, and declares it is all as it should be. Such are the benefits of civilization to the savage!

A little farther on, we fall in with a different sort of savage—an Irishman, on a little patch of ground which he cultivates after a fashion of his own, at the same time doing his housekeeping in preference to being "bothered with a woman." He is cooking his afternoon meal, which consists of a soup made from boiling a ham-bone, with thistles for greens, and a cup of spruce tea. Think of this, unlucky men, bothered with women, who, but for them, might be subsisting yourselves on thistles and spruce tea!

Our guide points out to us the peculiar features of Young's Bay, and the adjoining country. While we admire again the peaks of Castle (Saddle) Mountain, we listen to a legend, or tradition, which the Nehalem Indians relate of a vessel once cast ashore near the mouth of their river, the crew of which were saved, together with their private property, and a box which they carried ashore, and buried on Mount Neah-car-ny, with much care, leaving two swords placed on it in the form of a cross.

Another version is, that one of their own number was slain, and his bones laid on top of the box when it was buried. This, were it true, would more effectually keep away the Indians than all the swords in Spain.

The story sounds very well, and is firmly believed by the Indians, who can not be induced to go near the spot, because their ancestors were told by those who buried the box, that, should they ever go near it, they would provoke the wrath of the Great Spirit. The tale corresponds with that told by the Indians of the upper Columbia, who say that some shipwrecked men, one of whom was called Soto, lived two or three years with their tribe, and then left them to try to reach the Spanish countries overland. It is probable enough that a Spanish galleon may have gone ashore near the mouth of the Columbia, and it agrees with the character of the early explorers of that nation, that they should undertake to reach Mexico by land. That they never did, we feel sure, and give a sigh to their memory.

Some treasure-seekers have endeavored to find the hidden box, but without result. One enthusiast expressed it as his opinion, that he could go right to the spot where it is hidden; but why he did not do so, he failed to explain. Like the treasure of Captain Kidd, it would probably cost as much as it is worth to find it. Casting backward glances at the beautiful mountains, with their romantic foreground of forest and river, we turn toward Astoria. All along the edge of the wood which covers the point are hazel, wild cherry, alder, vine-maple, *spiræa*, mock-orange, and elder, besides several varieties of ferns, some of a great height.

Of the elder there are three varieties, all beautiful. The trees grow to a considerable size, and to a height of thirty feet. The colors of the berries are lavender, scarlet, and orange. We find also some other orange-colored berries, resembling immense raspberries, which our guide tells us are "salmon-berries." They are so juicy they will hardly bear handling, and literally melt in your mouth. Of the trees in sight, the most are fir, hemlock, cedar, and yew. But of whatever species are the trees, their unusual size and beauty make them interesting.

When we reach the point of the peninsula again— Point of Bowlders, we should call it—we are just in time to witness the golden changes of the sunset over Cape Hancock, and to see an ocean steamer coming in. She has passed Fort Stevens, and, by the time we have clambered over rocks and drift-wood to a smoother portion of the beach, is abreast of us, and almost within a stone's throw. We wave our handkerchiefs wildly, knowing, by experience, how pleasant is any signal from the land when our ship is coming in. Then, as if to answer us, she fires a gun, which stuns us with the report. We hasten to the wharf and scrutinize her passengers, while her captain exchanges courtesies with custom-house officers. In half an hour she is off again, leaving us to wonder how long it will be before Astoria gets her railroad, and ocean steamers discharge their cargoes within a dozen miles of the sea.

The situation of Astoria as a commercial *entrepot*, although, in some respects, a fine one, has its drawbacks, being cut off from the interior by the rugged and densely timbered mountains of the Coast Range; and, while it is true that the engineering science of the present day discovers obstacles only to overcome them, a good practical reason must be given capitalists for incurring enormous expenses. What course the railroad companies, now operating in Oregon, will pursue with regard to this point, can, at present, hardly be conjectured. The country now tributary to Astoria is a narrow strip of coast, which produces, like the Clatsop Plains, excellent vegetables, fruits, and dairy products, but is not usually well adapted to grain-raising. These products are continually increasing, as the numerous small valleys, in the radius of fifty miles, are being settled and improved; yet, it is our impression that the proper exports of this portion of the Columbia Valley are lumber, fish, and minerals, among the principal of which are coal and cement. The stone of which the new custom-house is built is taken from a quarry on the Washington side of the river, but is, by no means, handsome in color, or regular in stratification, being, apparently, formed from a deposit of sand around other bowlders, which are as hard as flint, and, occurring frequently, seriously interfere with the quarrying of regular blocks.

The Columbia, opposite Astoria, is six miles in width,

being one mile less than between the capes. The stage
of water on the bar, is, mean low water, twenty-four
feet; high water, thirty-two: from which it will be
seen that there is abundance of deep water, and room
for shipping, about Astoria.   About mid-river we had,
from the pilot-house of the *Grant*, one of the grandest
views to be obtained anywhere, of a magnificent body
of water, in conjunction with fine, bold scenery in
immediate connection, and distant visions of dazzling
snow-peaks.   Looking seaward, we beheld the dark
headland of Cape Disappointment, and the low neck
which constitutes Point Adams, with the broad opening of Young's Bay defining it more sharply; toward
the south, highlands, with Astoria at their foot, and
the "castled crags" of Saddle Mountain towering over
them; and toward the east, Mount Adams and Mount
St. Helen, each more than a hundred miles away, but
seeming to rise up in their pure whiteness out of the
everlasting green of the intermediate forests.

On the north side of the river, opposite Astoria,
we found the little fishing village of Chinook, where
salmon are yearly caught, and put up for export; and
the new settlement of Knappton, where is a fine lumber-mill, cutting about twenty-five thousand feet per
day, and where are also the cement-works, belonging
to the enterprising owners of the mill.   In a little
valley, just over the ridge back of this place, a colony
have lately settled, who pronounce the soil to be excellent, and themselves delighted with their situation,
especially as they are each entitled to a homestead of
one hundred and sixty acres of the choicest land they
can find.   But for lack of time we should have availed
ourself of the offer of our captain, and paid a visit to
the settlers of Deep River.

CHAPTER V.

AMONG THE FISHERIES.

HAVING satisfied ourselves that we have seen the principal points of interest about the oldest American settlement on the North-west Coast, we take passage, at an early hour of the morning, on board the *Dixie Thompson*, the elegant steamer which plies between Astoria and Portland.

Above Astoria, for some distance, there are no settlements on the river. But the grandeur of the wooded highlands, the frequently projecting cliffs covered with forest to their very edges, and embroidered and festooned with mosses, ferns, and vines, together with the far-stretching views of the broad Columbia, suffice to engage the admiring attention of the tourist. In consequence of fires, which every year spread through and destroy large tracts of timber, the mountains in many places present a desolated appearance, the naked trunks alone of the towering firs being left standing to decay. After a few years a new growth covers the ground, but the old trees remain unsightly objects still. It is true, however, in compensation for the ugliness of a burnt forest, that the shape of the country is thereby partially revealed, and that one discovers fine level benches of land fit for farming, in the openings thus made, where before no such variations from the general slope had been apparent.

The first point at which the river steamers touch in

going up, is Cathlamet—a small trading post and salmon fishery, about twenty miles above Astoria, on the north side of the river. Ten miles farther up, on the south side, is Westport, situated upon one of the numerous sloughs which the river forms on the Oregon side. This site was taken up as early as 1851, by Captain John West, who, with his family, has continued to reside here, giving his name to the place. Almost by his individual enterprise he has built up a flourishing settlement, and now owns wharves, warehouses, a store of general merchandise, a lumber-mill, and a salmon fishery, besides a fine farm and dairy.

This slough, or bayou, of the Columbia is a pretty bit of quiet water, with a level, wooded island on one side, and the main-land backed by wooded hills on the other. It is no place for a large town, but an excellent one for what it is—a flourishing trading post. The valley of the Nehalem, a considerable stream that runs nearly parallel with the Columbia, emptying into the ocean near Tillamook Head, is rapidly being settled up, and adds to the importance of Westport, which is the only trading post within twelve miles of the new settlement.

The steamer being detained for half an hour at this place, gives us an opportunity to step ashore and take a look at the salmon fishery. We find it a busy place, the fishing season, which begins in May and ends in August, being at its height. The manner of taking salmon in the Columbia is usually by drift nets, from twenty to a hundred fathoms long. The boats used by the fishermen are similar to the Whitehall boat. According to laws of their own, the men engaged in taking the fish, where the drift is large, allow each boat a stated time to go back and forth along the drift to hook up the salmon. The meshes of the nets are just of a size to catch the fish by the gills, when attempting to pass through; and their misfortune is betrayed to the watchful eye of the fisherman, by the bobbing of the corks on the surface of the water.

When brought to the fishery, they are piled up on long tables which project out over the water. Here stand Chinamen, two at each table, armed with long, sharp knives, who, with great celerity and skill, disembowel and behead the fresh arrivals, pushing the offal over the brink into the river at the same time. After cleaning, the fish are thrown into brine vats, where they remain from one to two days to undergo the necessary shrinkage, which is nearly one-half. They are then taken out, washed thoroughly, and packed down in barrels, with the proper quantity of salt. That they may keep perfectly well, it is necessary to heap them up in the barrels, and force them down with a screw press.

A *fishery* proper is understood to mean, a barreling establishment; while a *cannery*, is one where the fish are preserved in cans, both fresh and spiced, or pickled. The establishment of Mr. West is both these in one. This establishment, also, has commenced the business of saving the oil, which, in barreling salmon, is pressed out, and is equal in quality to the best sperm-oil.

The method of canning salmon was kept secret for one or two seasons, and only a few of the fisheries practiced it. No effort is now made to conceal the processes. The result is the main thing in which the public are interested, and this is a delicious preparation of fresh, or spiced and vinegared fish, put up ready for the table. The market for canned salmon is rapidly increasing — the principal exports being, at present, to California, South America, China, and the Islands. It is expected to find a market for it in New York and London, as soon as the amount produced becomes large enough to supply those cities.

The whistle of the *Dixie* warns us to bring our observation to a close at this point. Turning back down the slough, we emerge once more into the Columbia, and soon arrive at a point in the river known as the "Narrows," but to which Lieutenant Wilkes gave the name of St. Helen's Reach, from the bold view of that mountain obtained here, at a distance of eighty miles. The Narrows is a famous fishing ground, and the largest drift is here. Traps, or weirs, were also in use about the Narrows, but the high water, this year (1871), destroyed most of them. There are no less than seven fisheries in a distance of three miles, two of them being large establishments. That of Hapgood & Hume put up, this year, 700,000 pounds of canned salmon; West & Co., 400,000 pounds. Hume & Co., another firm, have also a large cannery, and Reed & Trott, another large establishment opposite these last, on the Oregon side. In all, there are twenty-five of these fisheries, from Chinook up to a point just above the Narrows, employing, altogether, about three hundred men.

The profits of the fishing business may be roughly computed by estimating the value of a case of canned salmon. An average salmon fills ten cans. These are put into cases containing forty-eight pounds each, and worth $9. Hapgood & Hume must then have put up, this year, over 14,583 cases, amounting to $131,247. About twenty men are employed about such an establishment during the fishing season, and eight or ten during the winter months. The winter's work consists in making barrels and cans. The cost of the labor of twenty men during four months, and of half that number during the remainder of the year, with the first cost of material, must be deducted from the total results, the remainder showing a handsome balance. And this is for only one cannery. Besides the two or three others, the different fisheries put up, this year, 2,000 barrels of fish.

The first drift for salmon catching was cleared in 1851, by Messrs. Hodgkins and Sanders—afterward continued by Hodgkins & Reed, now Reed & Trott—and the first canning establishment started, in 1867, by Hapgood & Hume. The buildings, erected at any of the fisheries, are of a rude character, being constructed of unplaned fir lumber. The largest are built about one hundred feet long, by twenty-five feet front, with a deep shed projecting over the river, for convenience in cleaning the fish as well as to shelter them from the sun. From the platform, extending along the side of the building, stairs run down to the water, where the boats are moored. In the lower story of this building are the vats, or "striking tubs," arranged around the sides. A commodious wharf, at which steamers and sailing vessels may receive freight, is also a necessary appendage.

There is no part of the Pacific Coast so well adapted to fish-curing as Oregon and Washington. The climate, either north or south of their latitude, is either too moist or too dry. Wood for barrels is close at hand; and, not yet utilized, close at hand, too, is the best salt in the world for curing meats of any kind. Seeing to what an immense business salmon fishing is growing, one can not help wishing that Nathaniel Wyeth, who tried so hard, in 1832, to establish a fishery on the Columbia, and failed through a combination of causes, could see his dream fulfilled, of making the Columbia famous for its fisheries and its lumber trade. But he, like most enthusiasts, was born too soon to behold the realization of the truths he felt convinced of.

There are several species of salmon and salmon-trout which are found in the Columbia. Of these, three species of the silvery spring salmon, known to naturalists as *Salmo quinnat*, *S. gairdneri*, and *S. paucidens*, are those used for commercial purposes, and known as the "square-tailed" and "white salmon"— the third species being considered as smaller individuals of the same kinds, though really distinct in kind.

When they enter the river, near its mouth, they may be caught by hook and bait. The Indians use small herring for bait, sinking it with a stone, and trolling, by paddling silently and occasionally jerking the line. Near the mouth of the Columbia they can be taken with the fly; but, as salmon do not feed, on their annual journey up the river to spawn, it is useless to offer them bait. They can only be caught at a distance from the ocean by nets and seines, or by spearing. The natives usually take them by using scoop-nets, which they dip into the water, at random, near the falls and rapids, where large numbers of salmon are collected to jump the falls. As these falls are all at a considerable distance from the sea, by the time they arrive at them the fish are more or less emaciated, from fasting and the exertion of stemming currents and climbing rapids, and, consequently, not in so good a condition as when caught near the sea. Hence, the superior quality of Chinook salmon.

The immense numbers of all kinds of salmon which ascend the Columbia annually, is something wonderful. They seem to be seeking quiet and safe places in which to deposit their spawn, and thousands of them never stop until they reach the great falls of the Snake River, more than six hundred miles from the sea; or, those of Clarke's Fork, a still greater distance. All the small tributaries of the Snake, Boise, Powder, Burnt, and Payette rivers swarm with them, in the months of September and October.

Great numbers of salmon die on having discharged their instinctive duty: some of them, evidently, because exhausted by their long journey, and others, apparently, because their term of life ends with arrival and spawning. Their six hundred miles of travel against the current, and exertion in overcoming rapids, or jumping falls, often deprives them of sight, and wears off their noses. Of course, all these mutilated individuals perish, besides very many others; so that the shores of the small lakes and tributaries of both branches of the Columbia are lined, in autumn, with dead and dying fish. But they leave their roe in the beds of these interior rivers, to replace them in their return to the sea by still greater numbers.

Besides the salmon of commerce, the Columbia furnishes a great many other species of edible fish, including salmon-trout, sturgeon, tom-cod, flounder, and smelt—all of which are excellent table-fish, in their proper seasons.

CHAPTER VI.

JUST above the Narrows, and opposite to the Oak Point of Captain Winship, is the modern Oak Point, which seems to have borrowed the name, and shifted it to the Washington side. The name is pretty and distinctive, and ought never to be changed, as it marks the western boundary of the oak - tree in Oregon and Washington. Between this and the sea not an oak - tree grows. The only business at or about Oak Point is that of the fisheries already mentioned, and the lumbering establishment of Mr. Abernethy, which was erected in 1848−9. It is run by water - power, and capable of manufacturing 4,000,000 feet annually.

About ten miles above Oak Point we come to the mouth of the Cowlitz River. Just below it is a high, conical hill, known as Mount Coffin. This eminence, together with Coffin Rock, seven miles above, on the Oregon side, formed the burial - places of the Indians of this vicinity, before the settlement of the country by whites. Here the dead were deposited in canoes, well wrapped up in mats or blankets, with their most valuable property beside them, and their domestic utensils hung upon the posts which supported their unique coffins. Wilkes relates in his journal, how his men accidentally set fire to the underbrush on Mount Coffin, causing a number of the canoes to be consumed, to the grief and horror of the Indians, who would have avenged the insult, had they not been convinced of its accidental occurrence. *Memelose Illihee* is the name which they gave to their burial grounds. Freely translated, it means *Spirit country*.

The Cowlitz is a small river, though navigable for twenty miles when the water is high enough, and about half that distance, at all times. It rises in Mount St. Helen, and runs, westwardly, for some distance, when it turns abruptly to the south. The valley of the Cowlitz is small, being not more than twenty miles long, and four or five wide. It is heavily timbered, except for a few miles above its mouth, where the rich alluvial bottom-lands are cleared and cultivated. No finer soil could possibly exist than this in the Cowlitz Valley. A few years ago, however, the town of Monticello, four miles from the Columbia, was all swept away in a flood. It has been replaced by a fresher edition of its former self, however, and looks as cheerful and ambitious as if it knew there could be no second deluge. Opposite Monticello is the old Insane Asylum for Washington Territory, in a location admirably adapted to confirm any incipient cases that may have appeared there. The asylum has recently been removed to Steilacoom, on the Sound— a very proper and delightful location.

This portion of the Cowlitz Valley does not depend alone upon its fertility for its future importance. There are extensive deposits of coal in the mountains which border the river, besides other mineral deposits which the North Pacific Railroad and an increase of population will eventually bring into notice. There is, too, an almost inexhaustible supply of the finest fir and cedar upon the mountains which hem it in.

The Cowlitz River, as might be conjectured, is a

rapid stream, and cold from the snows of St. Helen. Its waters in summer, when the snows are melting rapidly, are white, from being mixed with volcanic ashes, or some disintegrated infusorial marl or chalk. A favorite voyage for travelers coming down from Puget Sound, is twenty miles of canoe travel from Pumphrey's Landing to Monticello. An Indian canoe, with Indians to steer, carries one rapidly and pleasantly down stream—while the excitement of passing the rapids, and the splendid scenery of the wild, little river, furnish entertainment.

So disguised in a luxuriance of trees and shrubbery is the mouth of the Cowlitz, that, when we are in the open Columbia, we can scarcely detect the place of our exit from it. Crossing over to the Oregon side we find ourselves at Rainier, where lumber is manufactured, chiefly for export. The location of Rainier is, in many respects, fine; but, at present, there seems to be little besides the lumber trade to give it business, though there are a few excellent farms in the vicinity. Any day in summer one may see at this place a picturesque group of natives hanging about the wharves, or paddling their canoes near the steamboat-landing. Should they have berries to sell, they will offer them to you in neatly woven baskets of cedar-bark, which you are welcome to keep if you purchase their contents.

Without tarrying long, we steam on up, passing Coffin Rock—another *memelose illihee*—a promontory of basalt sparsely covered with trees, which have found soil enough in the crevices to support a stunted growth. Along here, on the Oregon side, is a tract of level land, extending back from the Columbia for some distance. It answers to the depression of the Cowlitz Valley; and it is remarkable, that, wherever a stream comes into the

Columbia large enough to be said to have a valley, there is on the opposite side a break in, or a curvature of, the highlands, making more or less level country facing the valley which is perpendicular to it, so that the valleys of the streams may be said to cross the Columbia, and, even, to be widest on the opposite side. Somewhere in here the Claskenine, a stream with a fertile and partially cultivated valley, enters the Columbia from the Oregon side; but the entrance is hidden by islands and shrubbery.

While we are interested in observing the stretch of the river at this point, and noting the islands and bayous which make it difficult to determine its actual breadth, we have advanced several miles, and find ourselves abreast of Kalama, the initial point of the North Pacific Railroad, on the Columbia River. Already an energetic beginning has been made, and from this port to the Sound a railroad will be constructed within a year or two. The silent grandeur of the Columbia is to be made busy and vocal with the stir of human labor, and the shriek of "resonant steam eagles" that speed from ocean to ocean, bearing the good-will of the nations of the world in bales of merchandise. It is the dream of Jefferson and Benton realized—only could the latter have had his wish fulfilled to live until this day!

"In conclusion I have to assure you, that the same spirit which has made me the friend of Oregon for thirty years—which led me to denounce the Joint Occupation Treaty the day it was made, and to oppose its renewal in 1828, and to labor for its abrogation until it was terminated; the same spirit which led me to reveal the grand destiny of Oregon in articles written in 1818, and to support every measure for her benefit since—this same spirit still animates me, and will continue to do so while I live—WHICH I HOPE WILL BE LONG ENOUGH TO SEE AN EMPORIUM OF ASIATIC COMMERCE AT THE MOUTH OF YOUR RIVER, AND A STREAM OF ASIATIC TRADE POURING INTO THE VALLEY OF THE MISSISSIPPI THROUGH THE CHANNEL OF OREGON."—*Letter of Benton to the People of Oregon, in* 1847.

But Benton did not understand the geography of the coast; neither did he know much of the practical working of railroads in recognizing or ignoring any points but their own. He did not foresee the Central Pacific going to San Francisco, and the Northern Pacific to Puget Sound, and an emporium of Asiatic commerce at either of these termini, while a third great city distributed their commerce along the Columbia and its tributaries, from its mouth to its sources; and that third city ought to be somewhere within a dozen miles of the present initial point of the North Pacific.

Turning this thought over in our mind, we are struck by the coincidence as some one points out to us, within the dozen miles, a place on the Oregon side which aspires to be that future city. It is a pretty town-site enough, certainly, sloping gently back from the river, which here, for two or three miles, has a smooth, gravelly beach, instead of the more usual abrupt and rocky shore. As we turn to the view of Mount St. Helen, just here seen through the canyon of the Cathlapootle, or Lewis River, which rises in the snows of that mountain, we agree that the aspiring town-site must command a beautiful prospect, including in its range Mount Adams and Mount Hood, as well as Mount St. Helen.

An admiring word calls out some volunteer remarks

from a fellow-passenger; and we ask, with augmented interest, what is claimed for this particular point. "In the first place," says our informant, "the Columbia River is the natural channel of commerce for the State of Oregon, as well as the southern border of Washington; for Idaho, and a portion of Montana. Its navigation is unobstructed from this point to the sea, which can not be said of it thirty miles farther up; besides, there are no good town-sites above the entrance of the lower Wallamet. The navigation of the river being easy, and safe for vessels of the largest size up to this point, is one good argument for us."

"Oh," we ejaculate, "you are interested in this place—what do you call it?"

"We call it Columbia City. Our view of the case," continues our informant, "is, that wherever the North Pacific Railroad has its crossing, there the greater portion of the domestic trade of Oregon will centre. The merchants of Eastern Oregon, Idaho, and Montana, in going to purchase goods, will not go by us, to San Francisco, or to the Sound, to purchase, if they can supply themselves just as well here, of which there can be no doubt. Direct importation by sea from New York, Canton, or the Islands, is just as easy here as to San Francisco, and only a few days longer from the first place. It is about two hundred miles nearer to China and Japan than any probable point on the Sound. It has to back it the great, fertile Wallamet Valley, and the country which contains it has fifty miles of river-front."

"All that sounds reasonable enough; but can the Columbia River compete with the Sound in the matter of safety? How is it about the bar?"

"There is not, nor ever has been, any more danger

on this bar than that of San Francisco or New York. Since the pilotage system was established, there has never been an accident on the bar. It is safer than navigating the Straits in a fog. There is no advantage in having more water than you can use, and there is enough and to spare in the Columbia. The Sound is 'the finest inland body of water in the world,' but you can not build a city all around it—there is nothing to support it. Talk about lumber and coal, and other minerals! Why, we have got the same here. Talk about ship-building and navy-yards, and all that! We can build ships, too; and we have the iron, within a few miles of us, to build into iron-clads, and *fresh* water for them to lie in. There's fifty to seventy feet of water right across the river at our point—and a mile wide at that!"

"Granting all you claim, that you could compete with San Francisco and the Sound—are not the Idaho and Montana merchants going to buy the bulk of their goods in Chicago?"

"Well, we hope to prevent that by judicious management. What we claim is, that the soil and population are going to fix the centres of commerce; and these we have on the south side of the Columbia."

There is so much common sense in this proposition that we refrain from contradicting it, and inquire the name of the little town with the beautiful location, at which the steamer is stopping. "St. Helen." A pretty name, and a pretty place; but why do the Oregonians repeat their names so much: Columbia River and Columbia City; Mount St. Helen and town St. Helen? Why not let every thing have a name of its own?

This is an attractive spot. The rocky bank forms a sharp, clear line of frontage, of a convenient height for wharves. A second bench, considerably more elevated, is covered with beautiful firs, in the midst of which stands a neat, white church. The village is grouped below, and has an air of cheerfulness not common to embryo towns. Our steamer is lying alongside the wharf of a lumber-mill, of a capacity evidently greater than any we have heretofore seen along the river. The mill is a fine structure, and the wharves are piled high with lumber, which is being loaded upon a vessel bound for Callao. There are several stores near the landing, and a whole fleet of little boats beached on a bit of sand close by. This is evidently a trading post of some consequence.

We take pains to inquire into the business and history of the place. Its history is a little peculiar. "Hope deferred which maketh the heart sick" has been its fortune from first to last. As long ago as when Wyeth was trying to establish American commerce on the Columbia, he selected this spot for his future city, and it obtained among the first settlers the name of "Wyeth's Rock." Afterward it was claimed by a man named Knighton, who, holding the same view of it, laid it out in a town-site, having it properly surveyed, the streets named, etc. But Mr. Knighton entertained such exalted notions of the value of his lots, and of his ability to build up a town without assistance, that those men who would have "stuck their stakes" in St. Helen, in a fit of pique, turned themselves into an opposition party, and laid out the town of Portland. By wiser management than Knighton's, they succeeded in drawing away from him the business he thought himself able to secure—and the result is, a city of ten thousand inhabitants at Portland, and only a couple of hundreds at St. Helen.

Such was the confidence in its future at the beginning, that the Pacific Mail Steamship Company built a wharf and warehouse here; stores and hotels sprang up; mills were built; and men were confident that their fortunes lay in this place. But, by and by, mysterious fires destroyed wharves, warehouse, and mill. The ocean steamer was forced to go to Portland; business died out; men became discouraged, and went otherwheres; and St. Helen was deserted by all except a faithful few, who never lost faith that time would bring all things right.

Six years ago the town-site changed hands, and the present large lumber-mill was erected by the St. Helen Milling Company, cutting from forty to seventy-five thousand feet in twenty-four hours. Two or three merchants set up general merchandising, and trade revived to such an extent as to rekindle hope in the hearts of the faithful few; and, now, St. Helen again asserts her claim to be considered "the best point on the Columbia River for a town." From all which it appears that Columbia City and St. Helen are rivals. As there is only a mile or two between them, it would not seem that their rivalry could be very fierce. Probably there will be, some time, an important town at or about one of these places.

St. Helen is the county-seat of Columbia County, and is situated at the junction of the lower Wallamet with the Columbia River. The country back of it, for about seven miles, is a series of benches, the first two or three of which are sparsely and picturesquely wooded, while the higher ones are well covered with timber. These benches are good farming and fruit lands, but not so fertile as the bottom-lands adjacent to the town-site—those of Sauvie's Island, and those on the opposite side of the Columbia—all of which country may be considered tributary to St. Helen, and, being well settled up, furnishes the present local trade of that place.

Scappoose Bay is a sort of bayou of the lower Wallamet, which sets back a distance of seven miles, and receives the waters of the Milton Creek—a fine water-power which might be turned upon the town-site of St. Helen, or made to furnish water-works for that place. There are, also, some fine grazing farms along Scappoose Bay on land subject to annual overflow.

Extensive beds of the richest iron ore lie adjacent to the township; coal exists in the mountains, six miles back; water-power and timber are plenty; while ships, of any size that can come into the Columbia, can lie alongside the natural wharves of trap-rock, that will keep off, forever, any encroachments which the river might make on a shore of sand. The views from the town-site are beautiful—from the bench, just back, magnificent. Game abounds in the vicinity: black bear, deer, grouse, partridges, and quail in the woods, and trout in the streams.

The country lying opposite St. Helen is the finest on the lower Columbia. The Cathlapootle, or, Lewis River, rises in Mount St. Helen, and, flowing south-westwardly, falls into the Columbia opposite the town of St. Helen. This river is a small and rapid stream, whose waters are as pure, cold, and clear as their mountain-springs. The valley of the main, or north, fork of the Cathlapootle is a rich, warm tract of country, producing excellent grain, fruit, vegetables, butter, and honey. It also raises stock for market, to a considerable extent. The road, or cattle-trail, from the Wallamet Valley to Puget Sound, passes up this

6

valley for some little distance. Annually, large numbers of cattle and sheep are driven to a market, on the Sound, by this trail, which, for want of a suitable ferry from St. Helen across, is not much used for wagons.

Another stream comes into the Columbia, within the sixteenth of a mile of the Cathlapootle. This is the Calapooya, or Lake River, which rises in a small lake near Vancouver, twenty-five miles to the east, and flows nearly parallel with the Columbia, until it empties into it. There is a large tract of excellent farming land along this river, also, most of which is already settled up. The farmers, from both these valleys, bring their produce to St. Helen to exchange for goods. The tide, at this point on the river, rises about four feet.

As we pass along up the Columbia from this point, we notice that the shores are level on both sides; for, here, within a distance of twenty miles, the Cathlapootle, Lake, and lower and upper Wallamet enter the great river. On the right is the fertile Sauvie's Island; on the left the bottom-lands, belonging equally to Lake and Columbia rivers—each shore densely wooded with cottonwood, ash, and willow, while, at a distance of several miles back, on either side, we behold the fir-clad highlands. This continues, without variation, to the head of Sauvie's Island, where a group of small islands, at the mouth of the Wallamet, give grace and variety to the river-view.

Passing the mouth of the Wallamet, we find that we are actually passing the foot of the Wallamet Valley, and that the flat country on the left extends all the way from the mouth of Lake River to the foot-hills of the Cascades; but, growing narrower as we near the mountains, is but the continuation of the Wallamet Valley into Washington Territory, according to the rule before noticed for the tributaries of the Columbia. Though this level country is now covered with timber, it must, from its alluvial nature, when cleared, prove very excellent farming land. That portion of it nearest the river is subject to the annual overflow; but there is no difficulty in determining the limits of submersion, for, wherever fir-trees are found, there the high-water never comes.

At a distance of about six miles above the Wallamet we come to the town of Vancouver, on the Washington side. This place is beautifully situated on a sloping plain, with a strip of velvety-looking meadow land on its river-front. It is the old head-quarters of the Hudson's Bay Company in Oregon, where resided, for more than twenty-five years, the Governor and Chief Factors of that company, nominally holding "joint possession," with the United States, of the whole Oregon Territory, but, really, for the greater portion of that time, holding it alone.

Here lived in bachelorhood, or with wives of Indian descent, a little colony of educated and refined men, who, by the conditions of their servitude to the London Company, were forced to lead a life of almost monastic seclusion. True, it happened sometimes that naturalists, adventurous travelers, and others drifted to this comfortable haven in the wilderness, and, by their talk, made a little variety for the recluses; and very hospitable they found them—ready to provide every civilized luxury their fort contained, without money and without price, so long as it suited them to remain.

There are few traces now of the old, stockaded fort.

When the British Company abandoned it, the United States Government took possession of it for a post; and, now, the traveler beholds scattered over the plain a town of a thousand inhabitants, and, bordering on it, the well-kept garrison grounds of the United States troops, with the neat officers' quarters encircling it.

Vancouver had, at one time, water enough alongside her fine wharves to accommodate large vessels easily; but, now, a sand-bar is said to be forming in front of the town, which is rapidly ruining her prospects of becoming an important river-port. There is, probably, no place along this low, alluvial land suited to the purposes of a large commerce. The changes likely to occur from the action of the annual flood on the sandy shores can hardly be calculated. Yet Vancouver must always remain the chief town of its county, and possess a good trade from the agricultural country back of it, which is already pretty well settled up, owning assessable property to the amount of a million of dollars.

Above Vancouver, for a distance of twenty miles, there are many beautiful situations all along on the Washington side, though the country is timbered heavily. The southern shore is lower: the Sandy—a stream coming down from Mount Hood—having its entrance into the Columbia above and opposite Vancouver, through alluvial, sandy bottoms. Beyond this the whole surface of the country becomes elevated, and we are among the foot-hills of the Cascade Mountains.

CHAPTER VII.

WE arrive now at what the tourist must ever regard as the most interesting portion of the river—the gorge of the Columbia. Here wonder, curiosity, and admiration combine to arouse sentiments of awe and delight in the beholder. Entering by the lower end of the gorge, we commence the passage, of fifty miles or more, directly through the solid mountain range of the Cascades. The snow-peaks, which looked so lofty at the distance of eighty miles, as we approach them gradually sink into the mountain mass, until we lose sight of them entirely. The river narrows, and the scenery grows more and more wild and magnificent.

Fantastic forms of rock—some with names by which they can be recognized—begin to attract our attention. Crow's Roost is a single, detached rock on the right, which time and weather are slowly wearing down to the "needle" shape, so common among the trappean formations. It stands with its feet in the river, at the extremity of a heavily wooded point; and in the crevices about its base, and half-way up, good-sized firs are growing. Above the Crow's Roost the mountains tower higher and higher. Frequently from lofty ledges and terraces of rock silvery water-falls are seen descending hundreds of feet, to some basin hidden by intervening curtains of wooded ridges. From the steamer's deck they look like mere ribbons; some of them, indeed, are dashed into invisible spray before they reach a level.

One of the handsomest of these falls has been named the Horse-tail, by somebody more given to ponies than to poetry. It has a straight descent, of several hundred feet, to a basin hidden from view, whence it descends by another fall to the level of the bottom-land, and forms another basin, or pool, among the dense growth of cottonwood, ash, and willow, which everywhere fringe the banks of the river.

Nearly opposite this fall is a high, precipitous wall of reddish rock, coming quite down to the river, and curving in a rounded face, so as to form a little bay above. This is the Cape Horn of the lower Columbia—a point where the Wind Spirit lies in wait for canoes and other small craft, keeping them weatherbound for days together. Fine as it is, steaming up the Columbia in July weather, there are times when storms of wind and sand make the voyage impossible to any but a steam-propelled vessel. It is at our peril that we invade the grand sanctuaries of Nature in her winter moods. The narrow channel of the river among the mountains, the height of the overhanging cliffs—which confine the wind as in a funnel—and the changes of temperature to which, even in summer, mountain localities are subject, make this a stormy passage at some periods of the year.

Sitting out upon the steamer's deck, of a summer morning, we are not much troubled with visions of storms: the scene is as peaceful as it is magnificent. Steaming ahead, straight into the heart of the mountains, each moment affords a fresh delight to the wondering senses. The panorama of grandeur and beauty seems endless. As we approach the lower end of the rapids, we find that at the left the heights recede and inclose a strip of level, sandy land, in the midst of which stands a solitary mountain (of basalt) called Castle Rock, about fourteen hundred feet in altitude. How it came there, is the question which the beholder first asks himself, but which, so far, has never been satisfactorily answered.

A mile or two beyond Castle Rock, situated on this bit of warm, sandy bottom-land, is the little mountain hamlet known as the Lower Cascades. Why it is that one name is made to serve for so many objects, in the same locality, must ever puzzle the tourist in Oregon. At the Cascades the tautology threatens to overwhelm us in perplexity. Not only is it the Cascade Range, which the cascades of the river cut in twain, but there are no less than three points on the north side, within a distance of six miles, known as the Lower, Middle, and Upper Cascades. Pretty as the name is, we weary of it when it is continually in our mouth.

It is a pretty spot, too, this Lower Cascades, surrounded by majestic mountains, and bordered by a foaming river; charmingly nestled in thickets of blossoming shrubbery, and can regale its guests on strawberries and mountain-trout. Here the Oregon Steam Navigation Company have a wharf and warehouse; and here we take our seats in the cars which transfer us to the Upper Cascades, and another steamer. We find the change agreeable, *as a change*, and enjoy intensely the glimpses of the rapids we are passing, and the wonderful luxuriance of vegetation on every side, coupled with the grandeur of the towering mountains.

At the Middle Cascades is a block-house, reminding us of the Indian war of 1855-6, and another one at the Upper Cascades. It is rare now to see an Indian at this point, where once they lived in large numbers, and had a famous fishing station ; and where, in still earlier times, they exacted toll from whoever passed that way.

The fall of the river in the five miles of rapids is about sixty feet ; but nowhere is there a perceptible fall of many feet together. The bed of the stream seems to be choked up with rocks, in such a manner as to suggest recent volcanic agency. At the Upper Cascades the river widens out again in a lake-like expanse, made picturesque with islands and handsomely wooded shores. In truth, all that portion of the Columbia, between the Upper Cascades and the Dalles, might very correctly be termed a lake—so little current has it, and so uniformly great is the depth of water—averaging forty feet, or twice the depth of the river below the rapids. From this fact, and that of the submergence of a belt of trees on either side of the river, for a long distance, the character of the hinderance to the flow of the Columbia may be very readily conjectured. At some period, long subsequent to the passage of the river through these mountains—a passage which evidently it forced for itself—by some violent means, a great quantity of rock was thrown into the bed of the stream, and, by forming a dam, raised the level of the water to its present height.

An effort has been made to secure the aid of Congress in removing this impediment to navigation. Great as would be the benefit, in a commercial point of view, of removing the dam at the Cascades, it presents itself unfavorably to the mind of the worshiper at Nature's shrines—one of whose happiest emotions must ever spring from the thought, that it is impossible for Man ever to intermeddle with the eternal majesty of scenes like these.

The material to be removed consists of a conglomerate of fragments of trap-rock, mixed with sand and earth. Embedded in this conglomerate are trunks of trees, often silicified—sometimes only carbonized, and sometimes both together. Of this silicified wood, there are many fragments to be found about the Cascades, embedded in the sand of the bottom-land. Of the trees standing submerged in the margin of the river, none of them are at all petrified; though, from the common occurrence of the fragments spoken of, the belief commonly obtains, that this is a petrified forest. The silica, which has entered into the pores of the silicified wood was, probably, derived from veins of that earth contained in the mass of conglomerate thrown into the river from the mountains at the time of the formation of the rapids.

From the deck of the steamer waiting for us at the end of the railroad portage, a beautiful picture is spread out on every side. The river seems a lake dotted with islands, with low shores, surrounded by mountain walls. Almost the first thing which strikes the eye is an immensely high and bold, perpendicular cliff of red rock, pointed at top with the regularity of a pyramid, and looking as if freshly split off from some other half which has totally disappeared. The freshly broken appearance of this cliff, so different from the worn and mossy faces of most of the rocks that border the river, suggested to the savage one of his legends concerning the formation of the Cascades: which is, that Mount Hood and Mount Adams had a quarrel, and took to throwing fire-stones at each other; and, with their rage and struggling, so shook the earth for many miles around, that a bridge of rock which spanned the river at this place was torn from its mountain abutments, and cast in fragments into the river. So closely does legend sometimes border on scientific fact!

While we are making this grave reflection upon the scientific truth of legends, some one presents us with a story, in rhyme, which he assures us is the true, original Indian legend of the formation of those other notable points on the river—the Dalles, Horse-tail Falls, Crow's Roost, as also the Falls of the Wallamet and Mount Hood. Making all due allowance for poetic license in some of the details, the story and the manner of its telling are worthy of notice; and we give it as a pleasing chapter of the early, romantic history of this romantic country.

THE SONG OF KAMIAKIN.

Should you ask me where I caught it—
Caught this flame and inspiration—
Should you ask me where I got it—
Got this old and true tradition—
I would answer, I would tell you:
Where the virgins of the forest
Sit with quills thrust through their noses,
Eating calmly cricket hashes;
Where the tar-head maid reposes;
Where the proud Columbia dashes,
Hearing nothing but his dashing.
Hias skookum (*) Kamiakin,
Of the vale of Klikatata—
Which I know each nook and track in
As well as Johnny knew his Daddy—
Was the chief of all the Siwash,
And the great high-cockalorem—
As his fathers were before him—

(*) Great-strong.

Of the winding Wallametta,
Which I sing—and say it surely
As the jingling Juniata
Sounds as well; but 'tis unpretty,
Poets of the sunset sea-rim
Flying off to Acropolis—
Very absurd it is, and silly—
While the glassy Umatilla,
And the classic Longus Thomas,
And the grassy Tuda-Willa,
All do flash and flow before us.

Well, my hero Kamiakin
Was in love; you know such folly
Must go in, or something's lacking
In all great, good rhymes emetic.
Now, she dwelt in Walla Walla;
But her Ma was awful stuck up;
And her pious Dad, ascetic,
'Gainst our hero got his back up;
And he swore on stacks of bibles,
Higher than the hay you stack up,
He would sue for breeches, libels;
He would sue him, shoot him, boot him—
That, in fact, he didn't suit him—
Didn't vote the proper ticket.

Now it cost him like the nation
Going from the land of cider
(You know how these Navigation
Fellows charge a horse and rider);
And, though he was law-abiding,
To be treated thus about her
He declared was rather binding,
And that he wouldn't go without her.
So he strode a cayuse charger
With white eyes, also white as
Foam of creamy, dreamy lager
From her nostrils to her caudle;
With a woolly sheepskin folding
Back behind his jockey saddle,
Where the girl could ride by holding.

Then while Dad on the piazza
Read the latest act of Andy,
And the maid on her piano
Trilled a ditty for some dandy,
"Chaco, chaco, cumtux mika?" (*)
From afar in tones coyote.
"Ah, you bet you, cumtux nika," (†)
Sang the maiden sotto voce
With this sign the chieftain sought her,
For the old man's bull-dog Touzer
Would have made it rather hot for
Kamiakin, Thane of Chowder.

Night and day they flew like arrows,
Till they passed by sweet Celilo:
"Bully," cried the chief; "tomollo's
Sun will see us hias lolo." (‡)
But the old man missed his daughter;
Vowing he would catch and score them,
Took the steamer, and by water
Reached the Dalles the day before them.

"Stop, you bummer," yelled the Daddy,
While the chief fled to the river;
And the Dad pursued, and had a
Henry-rifle, bow and quiver.
Then the chief wished him a beaver—
Big or little, didn't mind him—
But the gal, would you believe her,
Stuck like wax, tight on behind him.
Then she waved a wand of willow,
And behold the mighty river
(For the maiden was a fairy)
All did surge and shake and shiver,
Till the banks did kiss, or nearly,
And confine the foaming billow:
So they crossed without a ferry.

"Come back, come back, O Pickaninny—
Back across the stormy water,"

____

(*) Come, come—do you understand me?
(†) I understand you.   (‡) Far away.

Cried the old man, like a ninny.
One hand skewed her water-fall up,
While the other held her garter,
As they set off at a gallop.
O! she looked majestic, very,
As she answered, "Nary, nary!"
And the river so is flowing,
Though wider washed a foot or so,
For this was in the gleaming, glowing,
Gilded, golden long-ago.

Then they fled far down the river,
But the old man came upon them,
And she cried, "O Lord, deliver!"
And she blew a silver trumpet,
And she cried, "O hiac—jump it,"
Till the cayuse jumped the river—
Jumped the awful yawning chasms—
With the lovers both astride her.
Ah, enough to throw in spasms
Belles of this sweet land of cider!
But the Daddy, hot and snarling
At the chief and chieftain's darling,
Hip and thigh smote with his sabre,
While the cuitan was crossing,
And her silver tail was tossing;
And her long tail, white and shaggy,
Cleft, where Tam o' Shanter's Carlin
Caught the tail of faithful Maggie.

And that horse-tail still is flowing
From the dark rim of the river,
Drifting, shifting, flowing, going,
Like a veil or vision flurried,
But is never combed or curried,
As a body can diskiver.

"Verbum sat," now yelled the daughter,
As she with her lover vamosed;
And the Dad sat in the water
'Till he chilled and died, and so was
Turned to stone forever arter.
Now this Dad a noble Crow was,

And a chief of fame and power,
And is known unto this hour
As the "Crow-Rock" or the "Crow-Roost."

Well, they traveled in a canter
'Till they reached the sweet Wallamet,
And cried, "Boatman, do not tarry;
We will give three pound of salmon
If you'll row us o'er the ferry."
But he answered, "Nary, nary."
Then the maiden cried out, "Dam it,"
And the stream was dammed instanter

So the chieftain reached his nation,
And his mother gave a party—
Gave a July celebration—
And they dinnered very hearty,
All on kouse and salmon smoky,
And then danced the hoky-poky.

But her troubles grew the thicker,
As in truth so did the maiden,
For the chief began to lick her,
And distract her with upbraiding;
But she had to grin and bear it,
For the gods had got so mad, they
Said she never should repass the
Place she left her dear old Daddy
So she went up in the hill-tops
At the head of the Molalla,
For to look at Walla Walla;
And by magic spells and hoo-doo—
For, you know, she was a fairy—
She did manage soon to rear a
Mountain like the pile of Cheops.
And Siwash, who saw her mammuk, (*)
Called the peak "Old Mountain Hoo-doo."
But there came a Jewish peddler,
Packing head-gear, hoods, and "small tings"
(Says the Almanac McCormick).
And who didn't care three fardings

(*) Working or conjuring.

> For this dear and true tradition —
> As the learned like me and you do —
> And made the gross abbreviation
> Of Mt. Hood from Mountain Hoo-doo.

After this pleasant flight of imagination, we feel more than ever prepared to appreciate and enjoy our day's travel amid scenes so suggestive, only regretting that the author of the poem is unknown to us.

Badinage aside, the grandeur of the Columbia, for some miles above the Cascades, is so great and overpowering that one feels little disposed to attempt description. The Hudson, which has so long been the pride of America, is but the younger brother of the Columbia. Place a hundred *Dunderbergs* side by side, and you have some idea of these stupendous bluffs; double the height of the Palisades, and you can form an idea of these precipitous cliffs. Elevate the dwarfed evergreens of the Hudson highlands into firs and pines like these, and then you may compare. Considering the history, together with the scenery of this river, there is no other so complete in the impressions it conveys of grandeur.

Down this river, sixty-six years ago, floated those adventurous explorers, Lewis and Clarke. Seven years later the survivors of that part of the Astor expedition which came overland, were struggling along these wild mountain shores, among inhospitable tribes, trying to reach the fort at the mouth of the river. A few years later still, the "brigade" of the Hudson's Bay Company, annually, floated down from their hunting-grounds in the Rocky Mountains, jubilant at the prospect of soon reaching head-quarters—singing and dipping their oars in time, while their noisy gayety was echoed and re-echoed from these towering mountain walls.

Twenty-eight years ago, the first large immigration of actual settlers for Oregon came down from the Dalles in boats, furnished them by the Hudson's Bay Company, with much toil and danger, and some loss of life. To-day, we tourists gaze and dream at our leisure, from the deck of a first-class steamer, with all our wants anticipated. In another lustre, or in less time than that, the travel and trade of one-third of the continent may be borne upon this great highway of Nature, to and fro, between Orient and Occident.

But we have forgotten to observe the notable places. "This," says our Captain, "is Wind Mountain. The Indian name answers to our word *enchanted*, from the fact, probably, that when the wind is foul it is impossible to pass here with their canoes." On the south side, a few miles above the Cascades, is the beautiful place of Mr. Coe—a fruit farm among the foot-hills, and facing the Columbia. Here grow such delicious peaches as are rarely ever raised west of the mountains. A little settlement, at the foot of the mountains, is called Hood River, from being near the junction of that river with the Columbia. Opposite the mouth of Hood River a very fine view of Mount Hood is obtained. So near does it seem, that we see the glistening of the snow where its cliffs reflect the sun. Nearly opposite, the White Salmon enters, cold from the snows of Mount Adams, a glimpse of which we catch between the cleft heights of the river's gorge.

The farther we depart from the heart of the mountains the more marked is the change in the character and quantity of the timber. Firs have entirely disappeared, while spruce and pine have taken their places. The form, too, of the highlands is changed, being arranged in long ridges, either parallel with the river or at right angles to it, but all very extensive, and forming benches, dotted only with trees, instead of being heavily wooded, as on the western side of the range. The climate, also, is changed, and a dryness and warmth quite different from the western climate are observable.

More and more the basaltic formation constantly becomes visible, protruding from the hills on either side, and often appearing to wall in the river. Frequently it divides for a little space, leaving the prettiest natural slips for boats, and a clean, sandy beach, on which to make a landing; but only in a few instances have they been taken possession of, settlements along the river being rare. Occasionally, however, some hardy settler has taken up a farm on the narrow strip of alluvial land at the foot of the mountains; and doubtless a great many more might find homesteads in eligible situations along the river, where their nearness to market would enhance their value.

On nearing the Dalles the country opens out more and more, the terraced appearance continuing quite to that city, and the basalt here presenting a columnar formation. We come now to the last, and by far the most singular, portion of the gorge of the Columbia— the Dalles of the river. The river here flows for fifteen miles through a narrow channel, cut in solid traprock, and more or less tortuous. To eyes accustomed to the broad expanse of the lower Columbia, it is difficult to recognize the same river in the narrow, dark current that flows between walls of black, volcanic rock for so many miles above the Dalles. The river here not being navigable, by reason of its strong, swift current, its whirlpools and sunken rocks, we are forced to make our observations from the windows of the Oregon Steam Navigation Company's car, which makes the

7

portage to Celilo. The outlook, fortunately, is a good one; and we travel right along the river-bank nearly the whole distance.

What a strange scene it is! Sand, rock, and water—not uncommon elements in a pleasing picture; but here it is not pleasing—it is uncanny to a degree. We catch ourselves wondering how *deep* here must be a stream only forty yards wide, which in other places is two thousand yards wide, and deep enough to float any kind of a ship; for we can not help fancying that what the river here lacks in breadth it makes up in depth. But we are not aware that soundings have ever been taken in the Dalles.

Boats have gone through this passage. In low-water the barges of the Hudson's Bay Company used to run the Dalles. One or two steamers have been brought through at a low stage of water; but it is a very perilous undertaking—much more perilous than going over the Cascades at high-water. We make our observations, and conclude we should not like to take passage on this particular portion of the Columbia. How it swirls, how it twirls, how it eddies and boils; how it races and chases, how it leaps, how it toils; how one mile it rushes, and another it flows, as soft as a love-song sung "under the rose;" how in one place it seethes, in another is still, and as smooth as the flume of some sleepy old mill. A rock-entroughed torrent like none else, we pledge; and, in truth, is a river *set up on its edge.*

CHAPTER VIII.

FROM DALLES TO WALLULA.

DALLES CITY—or "The Dalles," as it is commonly called—is a town of about twelve hundred inhabitants, situated on the south side of the Columbia, at the lower end of the Dalles of the river. In the early history of the country it was fixed upon by the Methodists as a mission station ; but failing in their efforts to instruct the Indians, or intimidated by their warlike character, or both, they relinquished the station to the Presbyterians, who held it at the breaking out of the Cayuse war in 1847. On this occurrence the whole country east of the Cascades was abandoned by all missionaries of Protestant denominations, and Dalles was converted into a military station, the mission buildings having been burnt down.

When the Donation Act was passed, giving missions the ground previously occupied by them, the Methodists laid claim to a portion of Dalles. The Government, however, had appropriated a portion of the claim for a military post, paying for the part thus taken. The Presbyterians then disputed the claim, on the ground that they were in possession at the breaking out of the war, which compelled them to quit the place, and had never *abandoned* it, but had a right to return at the cessation of hostilities. The question of ownership has never yet, we believe, been satisfactorily settled.

The mining excitement, on the discovery of gold

in Idaho in 1862—3, first gave Dalles a start. In 1865 it was just such a place as one may see in any mining country—Nevada, for instance—a hastily built, rough-looking town, filled with restless, rough-looking men. The streets were dusty, there were no shade-trees, and very little comfort anywhere. Now, since the mining excitement is done away with, and only so much interest in it remains as a legitimate outfitting trade creates; and since the people here begin to understand the agricultural resources of the country immediately about them, Dalles has come to be quite a cheerful and handsome town. Real homes occupy the places of hastily erected board-houses; gardens blossom with exquisite flowers; shade-trees shelter and adorn the promenades; churches and school-houses abound; and the place is one of the pleasantest in Oregon.

The situation of Dalles is a fine one. Except in great floods like that of 1862 and 1871, the whole town is above high-water mark. It rises gradually back for a quarter of a mile, then sharply to a second well-defined bench of land, beyond which is a considerable ridge. The whole landscape back of, and surrounding, the town, is of fine outlines, and very handsomely ornamented with pine-trees.

A number of creeks fall into the Columbia, near Dalles City. Taking a ride up the little valley of Mill Creek, brought us through the garrison-grounds—a lovely spot—and out past some very pretty places and well-cultivated farms. It quite surprised us to come upon such well-to-do-seeming farmers, where the general aspect of the country is so uncultivated. But here is the evidence of successful and profitable farming: good houses, fine orchards, grain-fields, gardens, and fat cattle—the fattest and sleekest that ever we remember to have seen—sufficient proof of the nutritious qualities of "bunch-grass."

Just above the garrison-grounds is a beautiful view of Mount Adams, and another of Mount Hood. The little stream we are following up seems as if it came directly from the latter mountain, which does not look far off, but very real and solid, and near. We fancy that an hour's ride would take us up among the highest firs, quite to the glistening snow-fields; but it is forty miles away, still, with a very rough country between hither and yon, so that our hour would have to be lengthened to very many.

Chenoweth Creek, Three-mile Creek, and Five and Fifteen-mile Creek Valleys are all occupied by settlers. In every new country the first-comers choose the creek-bottoms and lowest valley-lands; especially in so dry a country as Eastern Oregon they have been considered of the greatest value. But farmers are commencing to experiment with wheat-growing on the uplands. To their own surprise they find the hills to be good grain-fields. Once the prejudice against the high-and-dry, rolling plains is done away with, there is no estimating the results; and yet we should say, on sight, that this country was only fit for grazing. So the fertile plains of California were once considered worthless for cultivation.

Wasco County, of which Dalles is the shire-town, extends along the Columbia River fully sixty miles, and toward the south nearly two hundred, covering an immense amount of territory ; and is drained by two rivers, of one hundred and fifty and two hundred miles in length. The whole population, probably, does not reach four thousand ; all those out of Dalles being either settlers on the small streams, or miners on the head-waters of John Day's River. Therefore Dalles has not yet much back country to sustain it. We are convinced, however, that in two or three years more a great change will have taken place in this respect, and that portions of Wasco County, hitherto entirely overlooked, will be made to "blossom as the rose."

A United States branch Mint had been partially constructed at Dalles, which was designed to coin the products of the mines of Montana, Idaho, and Eastern Oregon; but the opening of the Central Pacific Railroad, and the diversion of bullion to the Philadelphia Mint consequent upon it, have rendered a branch at Dalles superfluous; and the building will probably be converted to other purposes. A woolen mill has also lately been erected, which is to be supplied with material from the plains of Wasco County. A fine flouring mill manufactures a brand of "best Oregon;" the Oregon Steam Navigation Company have their machine-shops on a small island at the mouth of Mill Creek; and trades in general do a good business at this place. Churches and schools prosper among the Dalles people, and the population is rather more than ordinarily intelligent.

The name of Wascos was given to this division of the Des Chutes — so runs the Indian legend — in the following manner: The Indians being collected at the fishery, a favorite spot for taking salmon, about three miles from Winquat, one of them was so unlucky as to lose his squaw, the mother of his children, one of whom was yet only a babe. This babe would not be comforted, and the other children, being young, were clamorous for their mother. In this trying position, with these wailing little ones on his awkward masculine hands, the father was compelled to give up fishing and betake himself to amusing his babies. Many expedients having failed, he at length found that they were diverted by seeing him pick cavities in the rocks in the form of basins, which they could fill with water or pebbles, and accordingly, as many a patient mother does every day, adapted himself to the taste and capacities of his children, and made any number of basins they required. Wasco being the name of a kind of horn basin which is in use among the Des Chutes, his associates gave the name to this devoted father in ridicule of his domestic qualities; and afterward, when he had resolved to found a village at Winquat, and drew many of his people after him, they continued to call them all Wascos, or basins. To-day the tribe is little known, but the county of which Dalles is the metropolis bears the name once given in derision to a poor, perplexed father for descending to the office of basin-maker for his children.

The original Indian name of the place where Dalles stands was *Winquat*, signifying "surrounded by rocky cliffs." There are many Indian names attached to points in this neighborhood of poetical significations. "Alone in its beauty" is the translation of *Gai-galt-whe-la-leth*, the name of a fine spring near town. "The mountain denoting the sun's travel" is the meaning of *Shim-na-klath*, a high hill south of town, etc.

About three miles above Dalles is a noted fishery of the Indians, as mentioned above, and opposite to it is the site of the Indian village of *Wishram*, spoken of by the earliest writers on Oregon. No village exists there now—at least not any thing which could well be recognized as such. Like the ancient *Chinook*, it has dwindled to nothing.

Just opposite to Dalles is a handful of rather indifferent houses, constituting the village of Rockland, in the county of Klikitat, Washington Territory.

Aside from the river itself there is little to interest one between Dalles City and Celilo, the upper end of the gorge of the Columbia. There are rocks all about in every direction, a little grass, a great deal of sand, and some very brilliant flowers growing out of it. There are also a few Indian lodges, with salmon drying inside, whose rich orange color shows through the open doorway like a flame ; and a few Indians fishing with a net, their long black hair falling over their shoulders, and blowing into their eyes in a most inconvenient fashion. But every thing about an Indian's dress is inconvenient, except the ease with which it is put on! Some of these younger savages have ignored dressing altogether as a fatigue not to be undertaken, until with increasing years an increase of strength shall be arrived at.

The railroad takes us along under overhanging cliffs of plutonic rock, one of which is called Cape Horn, like its brother of the lower Columbia. As we near Celilo we discover that we have by no means left behind high banks and noble outlines. Just here, where we re-embark for the continuance of the up-river voyage, is a wide expanse of tumbling rapids, between lofty bluffs, rising precipitously from a narrow, sandy beach.

Of Celilo there is not much more than the immense warehouse of the Oregon Steam Navigation Company, nine hundred feet in length—built in the flush times of gold mining in the upper country—and the other buildings required by the company's business. Lying along the shores, in little coves, are numerous sailing craft of small size, which carry freight from point to point on the river above. The sun of an unclouded morning gilds their white sails, and sparkles in the dancing rapids. The meadow-lark's voice—loud, clear, and sweet—reaches us from the overhanging banks. It is at once a wild and a peaceful scene.

A short distance above Celilo the Des Chutes River empties into the Columbia, through a deep canyon. A remarkable feature of the rivers of Eastern Oregon is the depth of their beds below the surface of the country which borders them. The Des Chutes flows through a canyon in places more than a thousand feet deep. Where it enters the Columbia its banks are not so high, because the great river itself has its course through the lowest portions of the elevated plains; and its bed is nowhere at any very great elevation above the sea-level. At the Dalles, two hundred miles from the sea, the level of the river is one hundred and nineteen feet above it; and the Walla Walla Valley, at a distance of three hundred and fifty miles, has an elevation of a few feet over four hundred. Away from the Columbia, the elevation of the plains varies from five hundred to twenty-five hundred feet. Hence the great depth of the canyons of streams flowing on the same level with the great river.

Along this portion of the Columbia the traveler has plenty of time to conjecture the future of so remarkable a country—not being startled by constantly recurring wonders, as he might have been on the lower portion of the river. There certainly is great majesty and grace expressed in the lofty forms and noble outlines of the overhanging bluffs which border the river for great distances; and that is all. There is neither the smoothness of art, nor the wildness which rocks and trees impart to natural scenes; and the simple beauty of long, curving lines becomes monotonous. If it be summer, there are patches of color on the sere-looking, grassy heights ; rosy *clarkia*, blue *lupine*, and golden sunflower. We hear the voices of multitudes of meadow-larks ; and see a few prairie-hens stooping their long necks shyly among the bunch-grass ; or discover at long intervals a cabin, or a flat-boat, or a band of Indian ponies feeding.

We have leisure to study the peculiarities of this region : A great river, with a fertile country on either side of it, extending for hundreds of miles back, and having an annual "rise" as regular as that of the Nile. But this overflow does not affect the lands bordering upon it, because they are too high. What then? Is the country unproductive? No. It is a *dry*, but *not* a rainless country. Rain falls at intervals from September to June. Light snows cover the ground a portion of the winter season. The soil is of a mellow quality that does not bake with drought.

The first explorers of these high plains gave it as their opinion that trees would not grow below an elevation of two thousand feet, and that the lands adjoining the Columbia were only fit for grazing. This opinion, either borrowed from the early explorers, or suggested by the absence of trees in a wild state, was also held by the first settlers ; not only with regard to trees, but to all kinds of grain as well. There certainly could have been no more unpromising ground for the planting out of trees than that at Dalles. Yet, after four years of experiments, the streets of Dalles are lined with thrifty young shade-trees, and its gardens filled with fruit-bearing trees. Experiments with wheat have shown that it is not the bottom-lands alone which will produce crops, but the hills and ridges back from the rivers.

At Walla Walla—the lowest point near the centre of the Columbia River Plains—we are told that the same results have been obtained from experiments there. Five years ago Walla Walla was a seemingly barren spot; now its homes are embowered in shade from trees of a most astonishingly rapid growth. The wheat product of the Walla Walla Valley is no longer procured from its creek-bottoms alone, but farms are being laid out more and more among the rolling hills. Irrigation, where it can be made available, is resorted to ; but from what we have learned, we have great faith in the soil and climate to produce all that is necessary to man's support.

Civilization began in either hemisphere in the rainless countries of Egypt, Peru, and Mexico. The reason is evident. Civilization depends on the ease and security with which man harvests the fruits of his fields. The crop in the Nile Valley was unfailing, from the certainty and uniform duration of the Nile overflow. In Peru, from the constant presence of moisture eliminated from the atmosphere in the form of heavy dews, the cultivation of the earth repaid man's labor surely. On the high table-lands of Mexico irrigation was necessary, but once accomplished, there, too, agriculture flourished unfailingly ; and men, instead of roaming from place to place, settled and remained, until civilization arose and declined, by the natural processes of the growth and decay of nations.

In these countries, superior intelligence also resulted from the dryness of the climate; as it is well known a pure, dry air is stimulating to the mental faculties, while a moist, dull, or cloudy atmosphere is depressing. It is evident that men in a savage state, having the obstacles of want and ignorance to overcome, have been aided by these circumstances. Nor are they to be overlooked in considering the future of countries in the infancy of their development. The Columbia River Plains, owing to their elevation above the level of the draining streams, will probably require a system of irrigation by artesian wells, except those parts bordering on mountains whence water can be conducted with comparative ease. With this addition to the amount of moisture furnished by the light rains and occasional snows of winter, this great extent of country, now given up to the pasturage of Indian horses and a few bands of cattle, might be made to support a dense population, producing for them every grain and fruit of the temperate zone, in the highest perfection.

We are told, that when the missionaries went, in 1836, to look for a suitable place for a mission farm and station in the Walla Walla Valley, they estimated that there were about *ten acres* of cultivable ground within thirty miles of the Columbia River; and that was a piece of creek-bottom at the junction of a small stream with the Walla Walla River. These same explorers decided, that there were small patches of six or ten acres, in places, at the foot of the Blue Mountains, which might be farmed. As for the remainder of the country, it was a desert waste, whose alkaline properties made it unfit for any use. A few years' experience changed the estimate put upon the soil of the Walla Walla Valley; and now it is known to be one of the most fruitful portions of the Pacific Coast, and the quality of the soil really inexhaustible—its alkaline properties supplying the place of many expensive manures. And yet the capacity of the plains for cultivation has only just begun to be comprehended

Thirty-one miles above Dalles, we pass the mouth of the John Day River—a stream, in all respects, similar to the Des Chutes—with the same narrow valley, and the same depth below the general level of the country. What bottom-land there is along this river is already taken up, and there are mining-camps upon its head-waters, from which a steady gold product has been derived for the last eight years. The high bluffs intervening between the Columbia and the interior country quite conceal any appearances of settlement, and leave upon the mind the impression of an altogether uninhabited country—an impression quite erroneous in fact, though there are thousands of square miles still vacant.

Willow Creek is a small stream, coming into the Columbia thirty-three miles above John Day River, with a small, fertile valley well settled up. After an interval of another thirty-three miles, we find ourselves at Umatilla—a small town set in the sands at the mouth of the river of that name. It serves simply as a port to the mines of Eastern Oregon, and, as such, has a trade disproportionally large for its size. Here the steamers disembark their passengers and freight; and the stages and pack-trains take up what the steamer leaves, to convey it to the interior and the mines.

The Umatilla River, on account of its valley, is one of the most important streams of Eastern Oregon. The Umatilla Valley, together with the bottom-lands of several tributary creeks, furnishes a fine tract of rich, alluvial land, having a high reputation for its agricultural capacity. About seven thousand acres, nearly all bottom-land, are under cultivation in Umatilla County, the whole area of which is over forty-seven thousand square miles. Of course, large bodies of land are open for settlement; the variety of surface, in this county, ranging from mountainous and wooded to rolling prairie, covered with bunch-grass, and lastly, the narrow, but rich bottoms of streams, rendering it easy to select a farm or a timber claim, as may be preferred.

There is an Indian Reservation in the Umatilla Valley, where some farming is done by the Indians. Efforts have several times been made to have this reservation opened for settlement; but, probably, this will not be accomplished, as the Indians have no wish to sell. At an Agricultural Fair, held in this county, in 1864, the Indians took prizes on garden products. Indian corn, melons, vegetables, and fruits of all kinds attain great perfection in the Umatilla Valley, which is cultivable for a length of about sixty miles.

Thirty-five miles up the valley is the small, new town of Pendleton—the county-seat, beautifully located, and situated on the main lines of travel. A fine court-house is already erected by subscription of the citizens of the county. There is an excellent water-power in the vicinity, and every natural facility for convenient settlement. On the Indian Reservation, close by, are a saw-mill and grist-mill, and, in other parts of the county, six other mills manufacturing lumber; the timber for which is all procured from the forests of the Blue Mountains—the lumbering region of all this portion of Oregon, and of the south-eastern portion of Washington. There are about eight hundred square miles of timber belonging to Umatilla County.

The Indians on the reservation are the remnants of the Umatilla, Cayuse, and Walla Walla tribes, and number, altogether, less than one thousand. They.

are a partially civilized and peaceable people; yet whose presence as neighbors can not be particularly desirable. Their territory is unnecessarily large, amounting to a square mile to each individual.

All the way from the Cascade Mountains to Umatilla—a hundred miles, more or less—we have found the rivers all coming into the Columbia from the south side. Rising in the Blue Mountains, which traverse the eastern half of Oregon from north-east to southwest, they flow in nearly direct courses to the Columbia, showing thereby the greater elevation of the central portion of Eastern Oregon over the valley of the Columbia. At the junction of the Umatilla the Columbia makes a great bend, and flows nearly parallel with the Cascade Range instead of perpendicularly to it, receiving the rivers flowing east from the Cascades.

It is nearly sunset when the steamer quits Umatilla to finish the voyage we have entered upon, at Wallula—a distance of twenty-five miles farther up stream, in a direction a little east of north. We steam along in the rosy sunset and purple twilight, by which the hills are clothed in royal dyes. About eight in the evening we arrive at Wallula, too late to be aware of the waste of sand and gravel in which it is situated, and late enough to feel the need of rest; for albeit there is little enough of activity in steamer travel, there is plenty of fatigue, especially when one is sight-seeing, with the faculties of memory and attention continually on the alert.

Wallula is the port of Walla Walla Valley, and was long a post of the Hudson's Bay Company. As a site for a town, it has much to recommend it, in the way of beauty and convenient location; and, also, much to condemn it in the matter of high winds, sand,

and the total absence of vegetation. The bluffs bordering the Columbia at this place repeat those harmonics of grandeur with grace, which won remark
from us on other portions of the river. The Walla
Walla River, which comes in just here, is a very pretty
stream, with, however, very little bottom-land near
the Columbia.

The sand of Wallula is something to be dreaded.
It insinuates itself everywhere. You find it scattered
over the plate on which you are to dine; piled up in little hillocks in the corner of your wash-stand; dredged
over the pillows on which you thoughtlessly sink your
weary head, without stopping to shake them; setting
your teeth on edge with grit, everywhere. And this
ocean of sand extends several miles back from the
river, on the stage-road to Walla Walla, whither we
are going. Let us hope for such a merciful interposition as a shower!

CHAPTER IX.

If one does find the great billows of gravelly sand rather disagreeable to travel over, the first half-dozen miles out of Wallula, there is compensation in beholding the singular profusion of bright-tinted flowers that grow broadcast over the whole expanse—an example of the way in which the beautiful may flourish where the useful can not get a foothold.

A ride of three or four hours brings us to the crossing of the Touchet (pronounced *Too-shay*), the principal branch of the Walla Walla River. The course of the Walla Walla is nearly due west, and we are traveling in a parallel course toward the east. The Touchet has its rise in the same mountain of the Blue Range where the Walla Walla heads, but from the opposite, or north side of the butte, which is called Round Mountain. Its course is north-west, west, and south-west, to its junction with the Walla Walla, describing a semicircle, of which the Walla Walla is the base. In the same manner all the important tributaries of this river rise in Round Mountain, describe a lesser semicircle inside of the Touchet, and fall into the Walla Walla at about equal distances from each other: an arrangement by which this valley is exceedingly well watered; these creeks having other smaller ones tributary to them, and all flowing so near the surface of the ground as to be easily turned aside for purposes

8

of irrigation. The Touchet, at the crossing just mentioned, is a narrow stream flowing between banks of rich, black alluvium, with narrow bottoms of the same, covered with a tall, coarse rye-grass. A few farms have been commenced on the bottom-land, that look very lonely in so great a waste of uncultivated, perhaps uncultivable country; for we have left behind the sand and gravel, and come into a section where there are nothing but rolling hills of a light-colored earth, so fine and powdery that where the road has been used for a season, great canyons exist—the wheels of wagons and tread of animals having pulverized the soil, and the wind lifted and carried it away, leaving these deep cuts. This same ashen soil supports an abundant crop of the nutritious bunch-grass, and ought, therefore, under a system of irrigation, to be able to produce the cereals.

This ashen soil is certainly not pleasant to journey over in summer; and it is with real gratitude that the longed-for shower is welcomed, bringing an abatement of both dust and sunshine. A half-dozen miles from Walla Walla city, at the crossing of Dry Creek, the aspect of the country changes. Instead of rolling hills covered with bunch-grass, between the roots of which the gray earth is always visible, we come to a beautiful, level basin of land, bounded by the foot-hills of the Blue Mountains on the east, and stretching away off into undulating prairies on every other side. The first glimpse of this lovely valley is very cheering indeed.

Perhaps it is partly by contrast that the town of Walla Walla impresses itself so pleasantly upon the beholder on first entering it from the direction of Wallula. It does, at all events, surprise the traveler with its aspect of cheerfulness and thrift, with its neat residences and embowering trees, and the general air of comfort and stability which it presents. The population of Walla Walla city is in the neighborhood of fifteen hundred. Its trade is derived from a well-to-do farming community, and from outfitting for the mines of Idaho and Montana. Judging from the thronged appearance of the streets, the merchants are doing a profitable business.

Walla Walla enjoys the luxury of plenty of pure, bright, sparkling water. Mill Creek, one of the several semicircular streams before spoken of, passes through the town, and is diverted into a hundred tiny rivulets which course through its length and breadth, laughing, and glinting in and out every body's garden, carrying coolness, fertility, and music to each well-kept homestead. The flowers, vegetables, and fruits of Walla Walla gardens attest the use and the worth of these tiny canals.

In educational and religious institutions Walla Walla is very well represented. The Whitman Seminary was chartered in the winter of 1859–60, and built in 1867 by subscriptions from the people, to commemorate the services and sufferings of the Missionary martyr, Dr. Marcus Whitman, who, with his wife, and others, was murdered by the Cayuse Indians, in November, 1847. Out of this germ will probably grow the future University of the Walla Walla Valley. Aside from this institution, there are two high-schools in Walla Walla —one Protestant, and one Catholic; and a Teachers' Institute, organized in the summer of 1870, besides a number of public schools. Of churches there are several, well attended, and with flourishing memberships. There are two weekly newspapers, and a *Real Estate*

Gazette, published here. The city has two Masonic Lodges, one of Odd-Fellows, and one of Good Templars. It has telegraphic connection with Portland and San Francisco, and only needs a railroad to make it a young metropolis. Walla Walla, by the way, is the residence of Mr. Philip Ritz, whose intelligent efforts to get the Northern Pacific Railroad through this valley entitle him to the gratitude of its people.

The visitor to Walla Walla is expected to visit the site of the Waiilatpa Mission; and to one acquainted with its history this is not an uninteresting excursion. The "place of rye-grass" is the meaning of Waiilatpa, and just describes the point of bottom-land between the Walla Walla River and Mill Creek, near their junction. It could never have been a very cheerful place, being shut in by higher rolling prairie from any extended view; but was chosen according to the rules of all pioneers: water, and a piece of bottom-land. Besides, as mentioned elsewhere, the first explorers of the country did not understand it, and believed they had secured the only fertile spot when they settled on a low bit of creek-bottom.

Waiilatpa is just that—a creek-bottom—the creeks on either side of it fringed with trees; higher land shutting out the view in front; isolation and solitude the most striking features of the place. Yet here came a man and a woman to live and to labor among savages, when all the old Oregon Territory was an Indian country. Here stood the station erected by them: *adobe* houses, a mill, a school-house for the Indians, shops, and all the necessary appurtenances of an isolated settlement. Nothing remains to-day but mounds of earth, into which the *adobes* were dissolved by weather, after burning. Among the ruins are fragments of burnt glass, iron, earthenware, and charcoal—sole evidences that these heaps of earth are not like any mounds of the prairie round about.

A few rods away, on the side of a hill, is a different mound: the common grave of fourteen victims of savage superstition, jealousy, and wrath. It is roughly inclosed by a board fence, and has not a shrub or a flower to disguise its terrible significance. The most affecting reminders of wasted effort which remain on the old mission-grounds are the two or three apple-trees which escaped the general destruction, and the scarlet poppies, which are scattered broadcast through the creek-bottom,. near the houses. Sadly significant it is, that the flower whose evanescent bloom is the symbol of unenduring joys, should be the only tangible witness left of the womanly tastes and labors of the devoted Missionary who gave her life a sacrifice to ungrateful Indian savagery.

The place is occupied, at present, by one of Dr. Whitman's early friends and co-laborers, who claimed the mission-grounds under the Donation Act, and who was first and most active in founding the seminary to the memory of a Christian gentleman and martyr. On the identical spot where stood the Doctor's residence, now stands the more modern one of his friend; and he seems to take a melancholy pleasure in keeping in remembrance the events of that unhappy time, which threw a gloom over the whole territory west of the Rocky Mountains.

The Walla Walla Valley covers an area of eight thousand square miles, or 5,120,000 acres, and is contained within the boundaries of the Columbia River on the west, the Snake River on the north and east, and the Blue Mountains on the east and south. It occupies a

position nearly central with regard to the great plain
of the Columbia, and also the lowest point in it, being
only four hundred feet higher than the northern end
of the Wallamet Valley, or very little more above the
sea-level than the head of that valley, notwithstanding
the general difference in elevation between the country east and west of the Cascade Range. The greater
portion of the Walla Walla Valley belongs to Washington Territory; but a portion of it extends over into
Oregon, and touches upon the Umatilla country, which
it resembles.

The Government surveys have been extended over
820,000 acres. Of this, about 150,000 acres have been
taken up, and about 20,000 acres more of unsurveyed
lands settled upon. According to the census of 1870,
the whole number of acres improved in the county
of Walla Walla, in Washington Territory, is 63,377.
Its population is seven thousand; and the valuation
of property, real and personal, over three millions,
with no county indebtedness. The stock statistics of
1870 show 5,787 horses, 1,727 mules, 14,114 neat
cattle, 8,767 sheep, and 5,067 hogs. The cultivated
land was divided, for that year, as follows: Wheat,
9,561 acres; oats, 5,317; barley, 1,314; timothy,
1,522; corn, 2,795; besides smaller crops of rye,
buckwheat, etc. The number of fruit-trees in the
county, 60,525; flouring-mills, eight; saw-mills, four.
The grain crop, including wheat, Indian corn, barley,
oats, and rye, for the year 1871, or the crop now in
the ground, is estimated by the millers and others
competent to judge, at one million bushels; and the
fruit crop, for this year, much larger than ever before.
These figures give a very flattering idea of an interior
county only opened to settlement eleven years ago,

when the military prohibition, consequent on the Indian wars, was removed.

From a practical farmer and fruit-grower in the suburbs of Walla Walla city, we obtained the following estimate of the productiveness of the valley, and the season for harvesting grains and fruits:

AVERAGE YIELD PER ACRE.

25 bushels wheat.			*Fruit from seven-year-old trees and vines.*	
30	"	oats.		
30	"	barley	40,000 pounds apples.	
40	"	corn.	30,000 "	peaches.
20	"	rye.	40,000 "	pears.
40	"	peas.	50,000 "	plums.
36	"	beans.	20,000 "	cherries.
500	"	potatoes.	40,000 "	grapes.
200	"	sweet do	15,000 "	blackberries.
300	"	turnips.	15,000 "	raspberries.
1,000	"	carrots.	5,000 "	gooseberries.
800	"	parsnips.	10,000 "	currants.
20,000 pounds cabbage.				
$2\frac{1}{2}$ tons hay.				

YIELD OF EACH TREE, VINE, PLANT, AND SHRUB.

Bear from graft in three years.

	1st year.	2d year.	3d year.	4th year.
Apples	20 lbs.	50 lbs.	125 lbs.	250 lbs.
Peaches	15 "	35 "	100 "	200 "
Pears	20 "	50 "	125 "	250 "
Plums	20 "	50 "	125 "	250 "
Cherries	5 "	15 "	50 "	100 "

From off-shoot.

	1st year.	2d year.	3d year.	4th year.
Blackberries	3 lbs.	8 lbs.	15 lbs.	35 lbs.
Raspberries	3 "	10 "	20 "	40 "
Strawberries	..	$1\frac{1}{2}$ "	2 "	2 "
Grapes (at 2 years)	3 "	10 "	25 "	75 "
Gooseberries (at 2 y'rs)	2 "	5 "	10 "	20 "
Currants (at 2 years)	2 "	5 "	10 "	20 "
Pie-plants (at 2 years)	8 "	20 "	20 "	10 "

The various cereals and fruits of this valley are harvested as follows :

>Wheat, from the 24th of June to 10th of July.
>Oats, from 1st of July to 20th of July.
>Barley, from 20th of June to 1st of July.
>Rye, from 1st of July to 10th of July.
>Corn, from 20th of August to 10th of September.
>Strawberries, from 1st of May to 10th of June.
>Raspberries, from 10th of June to 20th of July
>Blackberries, from 25th of June to 1st of August.
>Gooseberries, from 20th of June to 1st of July.
>Cherries, from 20th of May to 1st of July.

The prices of labor range as follows :

Wages for farm laborers, per month.................... $35 00
Average wages of day laborers (without board)........ 2 50
Average wages of day laborers (with board)........... 1 50
Carpenters, per day 4 00
Female domestics, per-week 7 00

Number of farms in the county 654
Number of schools in the county 41

The climate of the Walla Walla Valley is, on the whole, a pleasant one. The average temperature is mild—about like that of Washington city; but less liable to sudden and disagreeable changes. The winters are occasionally severe; but, generally, snow does not remain on the ground longer than from four to ten days. Stock live and thrive, without care, on the natural pastures, both in summer and winter; yet, doubtless, some provision ought to be made against the possibility of a period of cold and snow. The spring and autumn weather is delightful, with light rains and occasional thunder-showers. The summer is warm and dry, with some windy days, when the dust is inconveniently omnipresent.

This is the climate not only of Walla Walla, but of nearly the whole eastern portion of Oregon and Washington — ranging from a mean winter temperature of thirty - five degrees to a mean summer temperature of seventy degrees at Dalles, and from forty - one degrees to seventy degrees at Walla Walla ; the winters being warmest at the latter place, and the summers about equal. Owing to the dryness and purity of the air, the atmosphere is never sultry, as in the Atlantic States, however warm it may be, and sun - strokes are almost unknown ; while both men and animals can endure to labor in the sun at a much higher temperature than in moister climates. Neither do violent wind-storms visit this country, such as are experienced in the Mississippi Valley ; nor earthquakes give people tremors, as in California. There have been one or two instances of the sudden rise of streams at the foot of the mountains, from a cloud - burst above, and the crops have been washed out of the soil by the sudden deluge. But these things are unusual, and are also a warning against building houses in narrow valleys, beside streams which head in the mountains. A higher location is at once more healthful and more safe, in any country.

We were invited out into the foot - hills, to visit a farm opened only the year before, where this accident by cloud - burst had destroyed the promising young orchard, garden, and a portion of the grain crop of the first year ; but owing to the fertility of the soil, and length of the growing season, the injury was, in a measure, repaired the same year. When we were there, during the first week in June, the rye in a field on the hill - side stood seven feet high, with occasional bunches several inches higher. The farmer—a young man from Iowa—was entirely satisfied with his new home, and was about to build a homestead on one of the sloping hill-sides of his farm, above high-water mark, from the sudden flood of the previous summer. We were also shown, at one of the farms, the fleece of a Cotswold sheep, with a staple thirteen inches in length, and glossy as silk.

There is no timber in the Walla Walla Valley except the cottonwood, birch, alder, and willow, which grow along the streams. The farmers are compelled to go to the Blue Mountains—generally a distance of fifteen miles—for timber for fencing, and lumber for building. Yet every farm is well fenced, and the farm-houses are better and neater-looking than those first erected in timbered countries, and for obvious reasons. The ugly, but substantial log-house, once erected, lasts a generation, and is tolerated from use. But where it is impossible to build such a dwelling, and where sawed lumber must be used, it is generally thought worth a little extra effort to put up something that the farmer will not want to tear down in his life-time.

A ride through the Walla Walla Valley, along the line of the stage-road, shows us the most cultivated portions, and a great deal of delightful country that is in its natural state. The face of the country is undulating—covered with grass and flowers. Fat, sleek-hided cattle feed in herds on a hundred hills. As we jog easily along over smooth roads, we enjoy the clear, bracing air, the cloudless sky, the glimpses of cultivation in wayside nooks, the flowers, the birds—the whole breezy, peaceful, harmonious landscape.

The only game which we notice is of the bird kind —prairie-hens and curlews. The latter amuse us much with their noisy, silly ways, and awkward style of running or flying. Thinking to learn something of natural history, we inquire of the driver of the mail (who, by the way, is from Maine) the use of the curlew's four or five inches of bill. The Yankee response is, "I don't know, unless it is to eat out of a bottle!" That reminds us to tell him about the man who became excessively fat eating mush and milk out of a jug with a knitting-needle.

Coming to the valley of Dry Creek we behold a new phase of this country. Dry Creek has bottom-land just wide enough, by intersecting it with transverse parallel lines, to make a row of farms extending its whole length. The views we catch of this winding belt of cultivation are perfectly charming—like the effects in a picture. Tints of green, yellow, and brown, in the fields; russets and grays; white houses in the midst of orchards and gardens; the beautiful forms of native and cultivated trees grouped about the houses, or fringing the creek; cattle, sheep, and fowls, giving life to the picture; or, better still, the farmer, with his children, coming in from the hay-field on the loaded wagon. While we gaze delighted, from every side the meadow-lark trills its liquid melody, in notes of exultation peculiarly infectious; and we find ourselves wondering why we have not always preferred the country to the town.

A ride of eighteen miles brings us to the Touchet; not the Touchet as we saw it at the first crossing on the road from Wallula, but a beautiful stream, with a gravel bottom, wooded banks, picturesque bluffs, and an open, handsome valley. And here, at the crossing, is the promising new town of Waitsburg. As the history of this place may serve as a hint to future pioneers in this country, we give it as it was told to us.

Six years ago a gentleman named Wait came here, in the winter season, and by dint of indefatigable exertion succeeded in erecting a flouring-mill before the next harvest. At that time there was no mill within a long distance, and the need of one was felt by the whole farming community of the Touchet. In less than two months after it was running, Mr. Wait's mill paid him five thousand dollars. Soon tradesmen of different kinds settled near the mill ; stores and a hotel followed, and in a short time a village had sprung up, which to-day has an appearance the most enterprising and thrifty of any town except Walla Walla in the whole valley. Judging by the farm-wagons, the sleek horses, the well-dressed farmers' families, and brisk trade at the stores, we should say that the Touchet was the farmer's land of Canaan.

Waitsburg has a school-house, which is its just pride, and that serves at present both as church and school building. There are sixty pupils in attendance, and a teacher of ability employed, at good wages, who also employs an assistant. Good morals and good order seem fashionable in Waitsburg—a great recommendation to a new place in a new country. There is considerable outfitting for the mines done at this place, which is on the direct road to Idaho.

A ride which we took out among the hills of the rolling prairie, convinced us that the bottom-lands were not the only grain-fields of Walla Walla Valley. Water is found by sinking wells to a reasonable depth ; and springs occur frequently in the ravines, from which water can be conducted, if needed, to irrigate lands on a lower level. We noticed several new farms, one or two years old, where there was the promise of future abundance and comfort; and here, as everywhere that the bunch-grass grows, we observe the fine looks of the stock subsisting entirely upon it.

Beyond Waitsburg the road follows along the Touchet Valley for twenty miles, past a constant succession of farms, with neat, commodious dwellings, and a neat, commodious, white-painted school-house, every few miles of the way. With such beginnings, the people of the Walla Walla Valley are on the high road to wealth and eminent social position, in the future of the State of Washington.

After leaving the Touchet, the road takes a course at right angles to all the streams, keeping up on the high ground except at the crossings. From the greatest elevations there are splendid views—wonderful for extent, and rather awful; inasmuch as we are able to realize that we are traveling like the fly on the orange, and can look down its slopes to dizzy descents of curvature. The crossings of the Tucanon and Alpowah rivers are any thing but agreeable coupons of travel. The hill of the Tucanon is frightful. Seeing the preparations made for the descent, prepares us for something of a hill; but when once started down the narrow, winding grade, with the coach seemingly minded to tumble over the backs of the horses, it is the most natural thing in the world to wish we had not undertaken the ride down. Walking, we reflect, if not an easy mode of locomotion, has the advantage of being eminently safe, compared to this. A mile and a quarter of such reflection prepares us to be thoroughly glad when the lowest level is reached, and we are in the little valley of the Tucanon, where again we find farms and pretty groves of glossy-leaved cottonwood. At the Alpowah we repeat the dizzying descent, with this difference, that once down we do not have to climb up again on the other side, but keep along its little valley to its junction with the Snake, when we have reached the extreme limits of the Walla Walla Valley on the north-east.

The Alpowah is a shallow, but unfailing stream, with a small, fertile border of bottom-land, cultivated chiefly by the Nez Perce Indians. The corn and melon-vines look unusually thrifty, and occasionally quite a comfortable house is to be seen; but, generally, a wigwam of matting, or a tent of skins, suffices for the requirements of these restless people. So near the Snake River the characteristics of Idaho begin to appear; excessive heat, and splendid flowering cacti, making gorgeous the hot sands of the riverside.

To sum up, before leaving it, the advantages of the Walla Walla Valley: we find that it is lovely in aspect, fertile, of a mild temperature, and well situated with regard to river and railroad transportation and markets, both east and west. The area of country upon the north-west coast, which will produce peaches, Indian corn, sugar-cane, sweet-potatoes, melons, and grapes, as well as the cereals, is limited, and confined almost exclusively to the territory east of the Cascades. Therefore, this valley has a double value, inasmuch as it will produce all these fruits, in addition to grains. It has, besides, innumerable facilities for manufacture, especially for woolen goods—the water-power and the wool being abundant. Nothing is lacking except railroad communications with the Columbia River and the East, to establish its importance; and that is what its citizens are now struggling to obtain.

CHAPTER X.

A GLIMPSE AT IDAHO AND WASHINGTON.

LEWISTON, in Idaho, where we find ourselves waiting for the semi-weekly steamer from the Dalles, is a place of only a few hundred inhabitants, situated on a sand-spit at the junction of the Clearwater with the Snake River, and between the two rivers. Immensely high bluffs of picturesque forms, bordering the northern shore of either, redeem the place from the appearance of insipidity which dead levels and barren sands ordinarily conspire to produce. In a business point of view, the location is a good one, whenever the development of the country, by means of settlement, shall demand a commercial centre. In flush mining times it was a lively place, being at the head of continuous navigation on the Columbia and Snake rivers. The most interesting of its institutions, to-day, is the depot for pack-trains, where miners' property is received, taken care of, and released to the owners upon the payment of certain dues.

Like Dalles and Walla Walla, Lewiston was at first considered hopeless as a soil for trees and flowers; but within three years past, their cultivation has been undertaken, with every prospect of gratifying success. We remark about Lewiston the same appropriation of high lands to wheat-growing that we have commented upon at Dalles. In fact, the highest level ground, in sight from Lewiston, is a table on top of the extremely high bluff on the north bank of the Clearwater; and this elevated plain is waving with green wheat-fields: that is, we take it for granted the wheat must be waving, up in that breezy locality—and waving is the conventional term for all cereals.

Lewiston is not without its notable resorts, of which a trout lake, about twenty miles away, is one; and many are the fishing parties who resort there to enjoy a catch and a basket-dinner.

We did not take time to visit the lake, but did take a ride out to Lapwai, the old mission station, and, more recently, military post and Indian Agency. The road to the Agency leads over the high prairie, where we find the tall grass—actually waving, this time—in the fresh breeze of morning, and very delightful it looks. The surface of the country about us is only slightly rolling, and covered with a bountiful crop of grass, which is rapidly being made into hay by a mowing-machine. Here, as everywhere we have traveled east of the Cascades, are the same varieties of flowers, in the same profusion; the same ever-present choir of sweet-throated larks; the prairie-hen, and grouse, and curlew. One other bird, the oriole, has its nest swinging from the branches of the cotton-woods in the vicinity of the Agency.

The little valley of the Lapwai is exceedingly pretty. The scene from the high prairie, before descending through the canyon of the creek, is in effect like a beautiful picture—with the garrison and the Agency, nestled each in its own nook, not far from each other. Lapwai Valley is very fertile, and, in early missionary times, was considered more productive than Waiilatpu in the Walla Walla Valley. When there was not grain enough at the latter station, it was brought from this place to eke out the supply. The old mission residence, now a ruin, stands near the bank of the Clearwater, close by the junction of the Lapwai Creek.

Here was the first printing-press ever used on this coast. It was a small hand-press, presented to the Oregon Mission, by the missionaries in the Sandwich Islands, and was used to print school-books and other works in the Nez Perce language. Great was the labor expended upon these efforts to enlighten dark minds; and poor the reward! Mrs. Spalding, who, with her husband, taught eleven years in this little valley, has long since passed to her rest, dying an early death in consequence of the shock to her heart and brain of the massacre at Waiilatpa, and the disappointment of her hopes. Mr. Spalding lives, but wrecked in health and spirits. And this is the average result of all missionary labor among the savages. The great error of the Government is in not making citizens, or subjects, of the Indians. Just so long as they are treated with as independent peoples, so long will Indian wars continue to exist.

The Agency is a quietly busy little place, with mills, offices, residences for the agent, interpreter, and others, a school-house, and last, not least, "Lyon's Folly"— a stone church, commenced by Governor Lyon during his gubernatorial term. The walls are standing uncovered, and will probably fall to ruin. Uncle Samuel must be a very good-natured relative to permit so many of his nephews to set up expensive monuments to themselves, and to pay themselves handsomely, at the same time, for doing it.

There never was a more stupendous piece of nonsense in the world than erecting handsome buildings, or providing any enlightened institutions for the use

of the average aborigine. The Nez Perces have always
been the exemplary "good Indians" of the North-
west; and they, certainly, are much better than their
southern neighbors of the same color; but to regard
them as civilized or half-civilized, or to expect them
to become such, is an error. We have forgotten what
the Nez Perce Treaty makes it necessary to expend in
educational facilities for the tribe; but we know that
there are about three thousand of these Indians, and
that there are, at present, *fifteen* in the school at the
Agency. It is true, that they cultivate a little ground
under superintendency, which looks well; but it is
only a little. They have an orchard, too, at the Agency,
but the fruit is all stolen while it is green, and never
does them any good. They parade themselves in their
blankets of red or white—lounging about, full of im-
pertinence, and very Indian altogether. Some of them
are fine-enough-looking fellows, and many of the young
women are pretty. The latter can learn to sew quite
nicely, but are too indolent to keep themselves decently
clad without constant urging. They prefer lounging
like the men, and amuse themselves in the Indian-
room of Mr. Whitman's, by chanting together their
low, lazy, not unmusical, though decidedly barbarous
and unpronounceable, sing-song.

The interpreter, an indispensable man at the Agency,
is Mr. Perrin Whitman, nephew of Dr. Marcus Whitman, of honorable memory. He has here a pleasant
home, and a cheerful family of his own; while the
Indians look upon him as the only person who can
represent them properly—therefore his position, probably, is a permanent one.

Asking to be introduced to "Lawyer," Mr. Whitman
took us to see this renowned chief. He is a rather

short, stout-built man, with a good face of the Indian type, very dark—almost African—in complexion, and dressed in a rusty suit of white men's clothes, with the inevitable high silk-hat. His manner, on being introduced, is a very good copy of the civilized man's; but his English is quite too imperfect for much conversation. We told him we had come a long way to see the man who had talked with Lewis and Clarke—at which he smiled in a gratified manner—when we asked him how old he was when Lewis and Clarke were in the country. He indicated with his hand the stature of a five-year-old child; but he must have been older than that, to have remembered all he claims to about the great explorers. It was his father, who, while they explored the Columbia to its mouth, kept their horses through the winter, and returned them in good condition in the spring.

On asking him the meaning of *koos-koos-kie*—the name Lewis and Clarke gave to the Clearwater—he explained, in Nez Perce, to Mr. Whitman, that Lewis and Clarke misapprehended the words of the Indians; that, on being questioned concerning this river, and knowing that it was the object of the explorers to find the great *River of the West*—as it was then called—they had answered them that the Clearwater was *koos-koos-kie:* that is, a smaller river, or branch only of the greater one beyond. But Lewis and Clarke understood them to give it as the name of the stream. "What was the name of this river, formerly?" we asked. He could not tell us. If it ever had a name it was forgotten; and thus, directly, the interview ended. It is remarkable, that so many of the rivers of the country are nameless among the Indians; and especially so, that the Columbia seems never to have had a name among any of the tribes residing either upon its shores, or in the interior.

Concerning the meaning of *Lapwai*, we were informed by Mr. Whitman that it meant the place of meeting, or boundary between two peoples, and that the Lapwai Creek really was the boundary between the Upper and Lower Nez Perces. The former tribe went to the buffalo-grounds, while the latter never did—hence the distinction. The habits of the two tribes were essentially different, as always are the habits of those who live by hunting from those who live by fishing and root-digging.

There is a double line of cottonwood and other trees from the Lapwai Creek to the site of the old mission mill, which at first we mistook for an avenue, but which turned out to be a spontaneous growth bordering the disused mill-race. The moisture furnished to the ground by the race, caused the flying seeds of the cottonwood to germinate where they fell, along its border; and the result is, a double row of fine, tall trees—a hint to the farmer who can turn water through his grounds from some spring or stream. Four or six years will find the trees grown to the height of thirty or forty feet.

On returning to Lewiston, we find the temperature to be about ninety degrees. The next day a sand-storm is prophesied, and really comes off. First comes the wind in full force, lifting the loose sands of the street roof-high, and forcing it into every aperture of the houses. Doors and windows are hastily closed, and people left to suffocate with heat, in preference to being suffocated with sand. Before night the clouds gather up, and give us at sunset a sharp thunder-shower—a common enough event in this eastern country, but almost unknown west of the Cascades. There is a considerable rain-fall during this night and another day—clearing and cooling the air—making our voyage down Snake River truly an agreeable one.

The impression gained of Idaho, from this glimpse at one spot on its western border, is more favorable than we had anticipated. It confirms the belief, which has been gradually growing in our mind ever since we left Walla Walla, that the whole of this Columbia Basin has been underestimated as a country for settlement. The people who are now here and have been here for several years, have always been in the habit of looking upon this as only a mining region. They have never intended to remain here longer than their interest in the mines compelled a residence, and, consequently, have never been inquisitive about its agricultural capacities. Within the last two years, however, a change has taken place in the sentiments of these unwilling settlers, and they are commencing to plant out fruit-trees—the surest evidence that they intend now to make this country their permanent residence. Several thousand dollars have lately been invested by the people of Idaho in young trees, imported from the East. In connection with these observations of our own, we clip the following from an Oregon paper:

"Mr. D. P. Thompson, the surveyor, was in town during the forepart of the week. In a conversation with him he informed us, that, in surveying north of Lewiston, this last summer, he was much surprised to find it such a beautiful agricultural country. He says, it contains more land adapted to agriculture than is embraced in the entire Wallamet Valley, including all its tributaries. He saw whole sections that would average one and a half tons of fine bunch-grass hay to the acre. The length of the valley is over one hundred and twenty-five miles, and contains millions of acres of land of a superior quality. This valley has now about seven hundred inhabitants, with new-comers daily arriving."

Doubtless, the same is true of many other portions of the Territory.

The Snake River below Lewiston is but a smaller copy of the Columbia above the Dalles—the same high, rounded bluffs, with frequent croppings of "eternal basalt," and the same high, rolling plains beyond. It has a current so rapid that the steamer, which has been thirty-six hours in coming up from Dalles, is able to return in fourteen. Leaving Lewiston at five o'clock in the morning, we pass no settlements, nor any streams of more consequence than the Alpowah and Tucanon on the south side, and the Pélouse on the north, until we arrive at the junction with the Columbia, at eleven o'clock, having traveled 149 miles in six hours. Eleven miles farther bring us to Wallula, where we left the steamer to take the overland route to Lewiston, through the Walla Walla Valley.

The notes which furnish the remainder of this chapter were imparted to us by several intelligent gentlemen of Walla Walla and Dalles, and we give them as furnishing the most reliable information concerning those remoter portions of the Columbia River Valley, which we had neither time nor opportunity to see with our own eyes.

Almost opposite the entrance of the Snake River into the Columbia, or, more properly, the junction of the north and south branches of the great river, the Yakima also joins itself to the Columbia. This is the principal tributary of the Columbia in South-eastern Washington; and, although the farming capacities of its valley are not yet very well known, it is believed that they are nearly equal to those of the Walla Walla Valley. The Yakima has its rise in a pass of the Cascade Mountains, from whence it flows south-eastwardly—receiving in its course numerous smaller tributaries descending like itself from the water-shed of the Cascades, and entering it from the west. Of these, the Wenass, Nachess, Athanam, and Pisco are the principal.

The Yakima Valley is the original home of the Indian tribe of that name, most of whom are now gathered on a reservation at Fort Simcoe, and, under faithful instruction, making some advance toward civilization. The number of settlers in the whole valley is about seven hundred, the most of whom are engaged in stock-raising. A direct trade with Puget Sound is carried on, to some extent, through the Snoqualmic Pass, as well as with Oregon, by wagons, to the Columbia River. Like Eastern Oregon this portion of Washington Territory is particularly adapted to stock-raising, whatever other resources it may ultimately develop. Its vast rolling plains furnish the most nutritious grass; its streams are frequent and pure, and bordered with cottonwood, alder, willow, and birch. Like the Walla Walla Valley it is destitute of timber—the material for fences and lumber all coming from the mountains, where yellow pine is found in abundance. The soil is a uniform light, sandy loam, with more or less alkali in it. Near the base of the mountains there is more loam and clay, and, as a consequence, the soil retains moisture longer than on the rolling plains. The river-bottoms consist in great part of rich alluvial deposits, which will cause them, ere long, to be turned into continuous stretches of farms, like the valleys of the Touchet and Dry Creek.

Such is the excellence of the bunch-grass peculiar to the plains east of the Cascade Mountains, that even the dry grass, which is cured by standing, keeps fat, all winter, the stock left to range at liberty anywhere on the prairie. Perhaps the pure, mild, dry, elastic nature of the atmosphere contributes something to keep animals in so good condition. Certain it is, that instead of coming out in the spring with lank sides and rough coats, they are as round and glossy as if kept up and curried.

At first thought it might be conjectured that such a country would be excellent for dairy purposes; but such is not the case. The dryness of the food and air together acts upon milch-cows to lessen the quantity of milk, although so much milk as is yielded is very rich in quality. Doubtless, many localities may be selected where dairies may be profitably conducted; but the tendency of cattle-raising is to a product of fat beef rather than butter and cheese.

The county of Yakima is bounded by the Yakima River on the east and north, the Cascade Mountains on the west, and the county of Klikitat on the south. Besides Walla Walla and Yakima counties, the whole of Eastern Washington is divided into Klikitat, with a population of two or three hundred, and Stevens, with a population of ten or twelve hundred. The former county borders on the Columbia, opposite Wasco County in Oregon, and consists almost entirely of high, rolling plains.

Stevens County, in the north-east corner of Washington Territory, contains 28,000 square miles. It is divided from south-west to north-east by the Clarke's Fork of the Columbia, the large and numerous branches of which furnish extensive tracts of fine agricultural valley-land. Colville Valley has been settled since the early times of the Hudson's Bay Company in Oregon, and was known, even then, to be a good wheat-growing country. In the Spokane Valley was a mission settlement as early as 1838, and would now have been a flourishing American settlement but for the hostility of the Indians, who, out of jealousy, forbade the cultivation of their grounds by the whites, until after the ratification of the treaties of late years. Until within the past year or two, the country was passed over only by miners going to, and returning from, the mines of Idaho and Montana. Now, there is a steady, though small, immigration of settlers into this county, especially in the south-eastern portion, bordering on Idaho, which is found to be a delightful country — good either for agriculture or grazing — consisting of large prairies of excellent land, interspersed with groves of timber, with abundance of pure water.

It would appear from these notes, that the best lands of Eastern Washington are not immediately along the Columbia, at any part of its whole course, but rather upon the upper portion of its tributaries, and upon tributaries of its tributaries, as the Walla, the Yakima, the Spokane, the Okinikane, and numerous other smaller streams, with their branches. The great plain of the Columbia occupies a central position with regard to these, and is a country in some parts worthless, and in others, fit only for grazing.

CHAPTER XI.

TRAVELING in Eastern Oregon is altogether the same as we have described it in the Walla Walla Valley, except that here there is more of it, and the roads at once better and worse for the same reason: that is, better graded on the hills, and more smotheringly dusty on the levels. Leaving Umatilla, where we arrive by steamer, there is the same sand-waste to toil through, and the same rolling plain of light, ashen soil to overcome, before reaching the settled portion of the valley, that there is between Wallula and Walla Walla. Nor is there any material difference between the general features of the Walla Walla and Umatilla valleys—their respective streams rising in the Blue Mountains, flowing in the same general westerly direction, and falling into the Columbia about twenty-five miles apart. By natural boundary, the Walla Walla Valley belongs to Oregon, lying as it does wholly south of the Snake River, and partly south of the Oregon line. As mentioned elsewhere, it is the lowest point in the Columbia Basin —the Umatilla, the Grand Ronde, and Powder River valleys being each successively more elevated than the other.

The whole extent of country, lying east of the Cascades in Oregon, is 58,000 square miles, and consists of immense plateaux, crossed from the north-east to the south-west by the Blue Mountains, from which numerous spurs put out in various directions. The best land in Eastern Oregon lies along near the base of this transverse chain of mountains, and in the valleys of the streams flowing from it on either side; the upper portion of these valleys being invariably the best. All the timber of the country—fir, pine, cedar, spruce, and larch—grows on the high mountain ridges, except the mere fringes of cottonwood and willow which border the streams. The Blue Mountains constitute a wall between the Columbia River Basin, to the north, and the Klamath Basin, to the south; hence all the rivers of Eastern Oregon head in these mountains, and flow into the Columbia and Snake rivers, only excepting those in the Klamath Basin, which empty into marshy lakes or sinks. Along these rivers, and about the lakes, there are large tracts of excellent land, suitable for farming. Subtracting from the whole area of Eastern Oregon what may be called the valley lands, the remainder is high, rolling prairie, with a considerable portion of waste, volcanic country in the central and western divisions. The country may be considered well watered throughout, as the streams are numerous, and water is to be found by stock at all seasons of the year. Owing, however, to the elevation of the plains above the beds of the principal streams, irrigation can not be effected, over a large portion of it, unless by artesian wells or by conducting water from the mountains. Such are the general features of that portion of Oregon lying east of the Cascade Mountains.

Attention was first drawn to the fertility of Eastern Oregon, by the population that rushed to the mines in 1861, and the three years immediately following. It became necessary to provide for the consumption of a large class of persons who dealt only in gold. The high prices they paid, and were willing to pay, for the necessary articles of subsistence, stimulated others to attempt the raising of grain and vegetables. The success which attended their efforts soon led to the taking up and cultivating of all the valley lands in the neighborhood of mines, and finally to experiments with grain-crops on the uplands, where also the farmers met with unexpected success. The nature of the soils on the south side of the Columbia is nearly identical with those already spoken of as characteristic of the north side: light, ashen, and often strongly alkaline, on the plains; sandy and clay loam at the base of the mountains, and richly alluvial in the bottoms, where it is often, too, mixed with alkali. It is discovered that on the highest uplands, and tops of ridges, there is a mixture of clay loam, which accounts for the manner in which wheat crops endure the natural dryness of the climate in the growing season.

Eastern Oregon has a population of about 13,000, and is divided into five counties, which serve for judicial purposes; but is more often spoken of by valleys than by counties. In one case, as in that of Umatilla, they are identical, where the county embraces this one valley. The reservation of the Cayuse, Walla Walla, and Umatilla Indians occupies a considerable portion of this county, which altogether has an area of about six thousand square miles—probably one-third. Of the remaining two-thirds, about half is reckoned as agricultural land, and the balance as grazing land of the very best quality. Water is plenty and excellent; but timber, as already described, is found only on the mountains.

Pendleton, the county-seat, is situated centrally, on

the Umatilla River, and is a thriving new town of two hundred and fifty inhabitants.   There are two or three other small towns in the county, each the centre of an agricultural district.   Two saw-mills manufacture all the lumber consumed in the county, which as yet has not more than 2,875 inhabitants, nor more than 8,000 acres of land under improvement.

Union County contains the valley of the Grand Ronde, a circular, grassy plain on the Grand Ronde River, long celebrated for its beauty and fertility. Here, in the early times of overland immigration by wagons, the weary immigrant found food for his cattle and rest for himself, after the long, exhausting march over the hot and sterile plains of Snake River.   This valley is thirty miles in diameter, well watered, and very productive in cereals, fruits, and vegetables, of all kinds common to the temperate zone.   About 15,000 acres are under cultivation in this valley.   The yield of grain-crops is unusually large, wheat often yielding from forty to sixty bushels per acre, and barley and oats, from sixty to eighty.   A considerable amount of land in this valley is subject to overflow, which makes it greatly esteemed as grass land, and for its annual product of hay.   Timber is conveniently near on the encircling mountains, and water abundant.

The climate of Grand Ronde Valley is subject to greater extremes than that of Walla Walla, or Umatilla, being nearly 1,000 feet higher than the latter. Snow seldom remains on the ground more than three weeks, the winter being short, and spring plowing and gardening commencing in March.   Although stock *should be* provided with shelter and food, yet cattle and sheep are often left to winter without either; and do very well without, in ordinary seasons.

Another fertile valley in Union County is the Wallowe, separated from Grand Ronde by a spur of the Blue Mountains. It contains about 6,000 acres of land similar to that in the larger valley. Eagle Creek Valley, of about the same extent as the Wallowe, contains also considerable good land; but is more celebrated for its mines than its agricultural advantages. The mineral resources of Union County are, in fact, important. Besides the gold mines, which have been profitably worked, there are indications of iron, copper, lead, and coal. Very few locations combine a greater number of advantages than this portion of Union County. Its mountains afford the precious and base metals, besides timber; its plains inexhaustible pasture; and its bottom-lands the most fertile farms. There are several hundred miners at work on Eagle Creek; and thirty or forty families settled in the valley.

La Grande is the county-seat of Union County, and contains 640 inhabitants. There are several smaller settlements, and altogether a population of 2,555. The grain product for 1870 was: Wheat, 250,000 bushels; oats, 200,000 bushels; barley, 150,000 bushels. It is estimated that its taxable property for 1871 is about $1,000,000. Four saw-mills supply the demand for lumber; as also do the flouring-mills the demand for flour. The stage-road from Umatilla to Boise City and the Central Pacific Railroad passes through La Grande, making communication easy with the Columbia River and the east.

Baker County, named for Col. E. D. Baker, who fell in battle at Ball's Bluff, embraces the valleys of Powder, Burnt, Malheur, and Owyhee rivers. Settled, like Union County, on account of its mines, it soon became well known for the productive character of its soil. We remember to have heard, while traveling on the stage from Umatilla, a miner from Powder River declaring, that "if a crow-bar should be left sticking in the ground overnight, it would be found in the morning to have sprouted tenpenny nails!"—after which assertion we never felt at liberty to question any statements which might be given us of the fertility of the Powder River Valley. With its several rivers, its bottom-lands, plains, and mountains, Baker County is one of the best in Eastern Oregon. Its elevation being four hundred feet greater than the Grand Ronde, gives it a climate both colder in winter and hotter in summer; the thermometer in winter sometimes falling to 15 degrees below zero, and in summer rising to 105 degrees. Yet the winters are short, and the spring early; while autumn is long and delightful, being a season of mildness and occasional refreshing showers.

Like Union County, Baker is celebrated for its mineral products. Placer gold has been found in considerable quantities in several districts, known as Rye Valley, Mormon Basin, Clark's Creek, Auburn, and Shasta. Later discoveries of rich gold and silver quartz-leads confirm its reputation as a mining region. Coal, and the base metals, are also known to exist here; the mining of which will be greatly facilitated by the presence of water and wood in abundance.

Baker City, with a population of 312, is the county-seat. The population of the whole county is roughly estimated at three thousand, but is probably less: as may be also that of other counties—since the estimate of the citizens seldom tallies with the general result of the census. It is very difficult to compute the shifting communities of mining counties accurately.

Baker County has several lumber-mills, and one flouring-mill; besides machine-shops, and all the necessary trades' and smiths' shops. Both in Union and Baker counties, great attention has been given to the establishment of schools. In the former, a common-school system is already in operation, supported both by subscription and taxation. Religious services are generally held with regularity, in the towns and settlements. In all that goes to make up the character of good, moral, and respectable people, the settlers of both Eastern Oregon and Washington appear to be ahead of most newly settled communities. The overland stage route is through this county, giving daily mail communication with the east and west. The roads are kept in good repair for wagoning goods from the Columbia River to the different mining-camps and settlements.

Grant County, embracing that central portion of Eastern Oregon where the Blue Mountains are highest, and extending southward to the southern boundary of the State, comprises altogether some of the most remarkable features of the whole country, including a portion of the wonderful "lake region." It was first settled in 1862, by a mining population; since which time it has contributed ten millions to the wealth of the world. The mining-camps are all on the headwaters and forks of the John Day River; where the placer mines are being worked out, only to be replaced by the discovery of rich quartz-leads.

This county, like those already mentioned, has become self-supporting, so far as farm products are concerned. It has under cultivation about nine thousand acres of land, chiefly on the North, South, and Middle Forks of John Day River, and a population of between two and three thousand. A good wagon-road from

Canyon City, the county-seat, to the Dalles, furnishes connection with the Columbia River; and excellent coaches, carrying a daily mail, travel over it. Freight-wagons and pack-trains also assist to keep the dry dust of summer stirring, tossing it to the boisterous winds that career at will over the boundless yellow plains of the Columbia.

In that portion of Grant County near the base of the Blue Mountains on the north side, it resembles in all respects those other mountain valleys already described, with its rich, level bottoms, grassy foot-hills, and timbered mountain ridges. But that portion of the county lying south of the Blue Mountains, is interesting not only as containing a large area of grazing and cultivable lands, but its physical conformation makes it a field of peculiar interest to the geologist. Harney Lake Valley, in this region, is remarkable for being a basin forty miles in diameter without an outlet. The lake from which it takes its name is a small, brackish body of water, near its south-eastern rim, receiving the drainage of the whole basin, and discharging it through some invisible outlet. Lying, as it does, in the most elevated portion of a broken and volcanic country, it affords speculation for the curious. The valley itself has a rocky surface, except in the northern part, where there is a tract of good arable land.

Besides Harney Lake, there is a chain of fresh-water lakes, commencing on the north-eastern, and extending to the south-western border of the county, in some cases connected by sluggish, but pure streams, and subject to high and low stages of water. They abound in fish and water-fowl, and are bordered generally by good grazing and agricultural lands; while

those bodies of water out of which no streams flow
are all more or less alkaline, from receiving the drainage of the alkaline soil about them, and not discharging any portion of it.

Wasco County, extending from the northern to the
southern boundary of Oregon, along the base of the
Cascade Range, and having a breadth of more than
sixty miles, comprises almost every variety of surface
and soil belonging to all the other counties. Its
southern portion, like the southern portion of Grant
County, is a lake country. A chain of volcanic highlands, commencing at Diamond Peak of the Cascade
Range, runs north-easterly, joining on to the Blue
Mountains, and separating this lake-region from the
valleys of the Des Chutes, Crooked, and John Day
rivers, which flow toward the north; making of this
south-western portion of Eastern Oregon an isolated,
as it is a peculiar country.

Lying near the base of the Cascades, and south of
the ridge just mentioned, is the Klamath Marsh, a wet,
grassy basin, out of which flows Williamson's River, a
stream of considerable size, into Great Klamath Lake,
a few miles farther south. Near the head of this lake
is situated Fort Klamath, a military post, located here
during the disturbances with the Snake and Klamath
Indians in 1863. On the eastern shore of this lake is
located the reservation of the Klamath, Modoc, and
Snake Indians. It occupies a tract about fifty miles
square, including the marsh and the connecting river.
The general appearance of the country which the reservation embraces is sterile and volcanic. In shape it
is rolling, covered with a fine growth of yellow and
sugar pine, with some cedars, firs, and on the streams,
cottonwood, poplar, and willow. The best part of the

reservation is that which lies on Sprague River—a stream rising about forty miles to the east, in the highlands about Goose Lake, and flowing westwardly into Klamath Lake. This valley, fifteen miles in breadth by forty in length, possesses a quick, fertile soil; although its elevation of four thousand feet above the sea, unfits it as a region for the farming of the tender fruits and vegetables. Wild flax grows abundantly in this region, as it does also in many parts of Eastern Oregon.

Springs of pure, clear, cold water are very numerous; some of them of immense size. There is one bursting out at the base of the mountains about two miles west of Williamson's River, which is a quarter of a mile across in one direction, and twenty-five rods in the other, and which discharges a stream of clear, cold water large enough to be navigable by the steamers that run on the Wallamet River. This water, flowing into Williamson's River, completely changes its character, from warm and turbid to clear and cold; in which trout from twelve to sixteen inches may be plainly seen disporting themselves at a distance of several yards. [A spring of a similar character and dimensions bursts out at the foot of the Cascades, a few miles north of the Three Sisters, discharging itself into the Des Chutes River.] The saw-mill at the Agency is run by the same spring which supplies an irrigating ditch; and has besides a large surplus, a portion of which will be used in running a grist-mill. The lands of the reservation, however, that have been put under cultivation, are too high and too cold ever to produce the farming results to make it self-supporting. Game, fish, and roots, such as the Indians use, are abundant; and on these the Indians can at least partially subsist themselves, while being taught to labor.

South-east of the reservation, and beyond the western ridge of the Goose Lake Mountains, is Goose Lake Valley, containing a considerable portion of good agricultural land, with a much larger amount of excellent grazing land. Surprise Valley, on the eastern side of Goose Lake, is similar to those previously mentioned; and all are surrounded by timbered ridges. Goose Lake and Surprise valleys are well settled up. There is, in fact, a succession of settlements lodged in the small valleys of this portion of Oregon, all along the California and Oregon line. It is estimated that ten thousand head of cattle are pastured on the meadows about Clear Lake—a country hardly known as yet.

The Oregon Central Military Road passes through all this region, starting from Eugene City in the Wallamet Valley, and crossing the Cascades at Diamond Peak Pass. From thence it crosses to Owyhee, in Idaho; passing through much valuable mineral country also. The road from Chico, in California, to Boise City, in Idaho, traverses the south-eastern corner of Oregon; a great deal of freight going that way to the mines. It is hoped to bring a branch of the Central Pacific from the Humboldt Valley, through the Klamath Lake region, into the head of the Wallamet Valley. A scheme is on hand for turning the waters of Lost River, a stream which flows out of Wright Lake into Clear Lake, through a canal, which shall conduct them into the Lower Klamath Lake; thus draining thousands of acres of excellent land, well adapted to settlement.

The northern part of Wasco County contains the valleys of the Des Chutes, John Day, and Crooked rivers, and their tributaries; besides the valleys of numerous creeks falling into the Columbia, near the Dalles—all of which are pretty well settled up. Proceeding north from the Klamath Lakes, we first come into a country interesting chiefly to the geologist; being an immense plain covered with volcanic ashes and tufa, except a small valley of good land on the head-waters of the Des Chutes, in the vicinity of a cluster of beautiful lakes. North of these are the Three Sisters—a beautiful group of snow-peaks, standing out from the range, and covered with snow almost to their bases. For a long distance to the east of these, the country is a waste of volcanic ashes and cinders, into which the legs of a horse sink eighteen or twenty inches. In the midst of this waste is an old crater of a volcano, its walls still standing to a height of between two and three hundred feet; and in its neighborhood lava, scoria, and obsidian are scattered broadcast. About the sources of the Crooked River, an affluent of the Des Chutes, are also numerous boiling springs, indicating the volcanic nature of the country.

Passing the spring before mentioned as discharging into the Des Chutes, and crossing two or three small streams of clear water, cold from the snows of Mount Jefferson, we come to the Warm Springs Reservation, the home of the Des Chutes, Wascopams, and several other tribes of Indians. The reservation takes its name from the boiling springs in its neighborhood, which are curiously near to a stream of ice-cold water. The country here is high, and worthless, except for grazing; and can never be made to support the Indians gathered upon it. In the vicinity of this reservation a bed of moss-agates has lately been discovered which promises to be quite extensive.

Following down the Des Chutes, we cross several creeks coming into it. One of these, Tyghe Creek, falls into this river at a point where the canyon it flows through is more than a thousand feet in depth. From Tyghe Creek it is thirty-five miles to the Columbia. Not far below the entrance of the creek, the road from Dalles to Canyon City crosses the Des Chutes. While here we are almost abreast of Mount Hood, and spread out on every hand is a landscape of wonderful impressiveness and extent.

Dalles is the county-seat of Wasco County. Its population is between 700 and 800, while that of the county is 2,489. The proportion of its urban to its suburban population shows the greater number of people engaged in agriculture and stock-raising; for Wasco County has no towns except Dalles. There are four saw-mills in this county, and one large flouring-mill at Dalles; a woolen-mill—not in operation at present; and extensive machine-shops, as mentioned elsewhere.

A railroad is projected, to begin at a point on the Union Pacific, and following up Ham's Fork of Green River, and along Bear River to the nearest point on the Snake River; to follow the Snake Valley down to the Immigrant-crossing; through the mining counties of Eastern Oregon, and so on to Dalles City, on the Columbia. There are many arguments in favor of this route to the great river thoroughfare of Oregon. Such a road would inevitably be continued to the Wallamet Valley, and form connection with the Northern Pacific to Puget Sound. Cheap transportation is the great want of the whole upper country. High prices for labor and for all the commodities of life must prevail, when $30, coin, is the price of transporting one ton of freight, by measurement, from Portland to Umatilla. . Add to this the freight from San Francisco, seven dollars, and the additional freight by wagon to places remote from the Columbia, and goods become worth "their weight in gold."

Even at these figures, there is landed at Umatilla, annually, fifteen to twenty thousand tons; and at Wallula, in Washington, five or six thousand more. As a consequence, wages range from six to eight dollars per day. It is proof positive of the worth of the country that it continues to grow and prosper under such disadvantages. The surplus of grain which is raised will not, in most cases, pay for shipping to foreign markets. One example to the contrary, however, came under our observation, where Mr. Wait, of Waitsburg, in the Walla Walla Valley, shipped several thousand barrels of flour to Europe and made a dollar a barrel on it; and this year, as much as 100,000 bushels of wheat were exported from the Walla Walla Valley, by way of the Columbia River. If, then, with so much against him, the business man can make money, how many times would his chances be doubled by quick and cheap transportation. Railroads are truly the one great need of all this country, and with them would come the population to make them paying.

Having seen enough of Eastern Oregon, on our return to Dalles we take steamer for the Lower Columbia, and Wallamet River. We rise early, as any one must who goes anywhere in Oregon, and get our last and loveliest view of Mount Hood from the east side of the Cascades. We have seen it in every possible conjuncture of circumstances, almost at its feet, and where distance made it seem like a faint white cloud. But behold his majesty this morning, draped from summit to base in a golden-tinted tissue of morning mist, of so delicate a texture that it has no trait of masculinity about it! In fact, we are reminded of a girl in the " trying on" process with a straw-colored silk grenadine. Her head has not yet emerged from the billows of gauze; nor is her robe quite shaken down on one side—the shining petticoat of snow showing daintily underneath. Many are the masquerading costumes and airs of the solid old mountain, who, despite the dignity of his thousands of years, affects at times the blushes of the rose.

CHAPTER XII.

THE upper mouth of the Wallamet is about eighteen or twenty miles above the lower one—the Wallamet and Columbia being separated for this distance only by a narrow, sandy island, which in the period of the summer flood is two-thirds under water. The original name of this island (now called Sauvie's) was Wappatoo, from the abundance of a plant of that name (*Sagittaria sagittifolia*) found growing there. This plant has a tuberous root, which is used by the Indians for food, and grows most abundantly in marshy places or shallow lakes. "Wappatoo lakes" are also considered valuable fields for fattening hogs; and the interior of Sauvie's Island furnishes several of these. Notwithstanding that every summer their farms are under water from two to three weeks, most small farmers prefer the quick, warm soil of the island, to higher ground on the mainland. Here, after the freshet subsides, about the second week in July, crops of potatoes, melons, and vegetables may be put in, and come to maturity quite as early as if planted in the spring, on the colder soil of the uplands. Sauvie's Island is indeed the garden of the lower portion of the Wallamet Valley.

The upper mouth of the Wallamet comes out between the head of the island, and a low, sandy point opposite. From the formation of the land it appears probable that the island was once a continuation of this point—a peninsula, in fact, which was finally cut off by some heavy winter flood forcing its way over and through it. Numerous small islands form quite an archipelago above the head of the principal island. As they are all densely wooded with willow, cotton-wood, and other water-loving trees, they present a very picturesque view.

Supposing ourselves to be standing on the hurricane deck of a steamer, passing among these islands, between the Columbia and the Wallamet, with stretches of both in sight; with the heavily wooded shores of both rivers plainly visible; with the Cascade Range drawn in blue on the eastern horizon, and the white peaks of St. Helen, Hood, Adams, and Jefferson rising sharp above it; and over all a rosy, sunset sky, its reflection coloring the rivers and tinting the snow-peaks—we would hardly expect ever to meet a lovelier picture than this one before us.

The Wallamet River, unlike the majestic Columbia, divides, nearly in half, a level valley of open prairie-land. Hence, and because the earliest settlers of a country always select the lands easiest of tillage, we find nearly the whole of the Oregon population in the Wallamet Valley. Had we entered by the lower mouth, and come up on the south side of Sauvie's Island, we should have found the land on either side divided into farms, and have witnessed the shipping of stock, and other signs of local trade; although here the valley is limited to a plain of half a mile to a mile and a half in breadth, bounded by a ridge of high, fir-clad hills.

From the head of the island up to Portland, a distance of little more than six miles, the hills continue to follow the southern bank of the river at about the same distance back; while the opposite bank is only moderately high, and rolling. We pass by Springville, a grain depot, on the right-hand side; and St. John, a stave factory, and small settlement, on the other. Farm-houses grow more frequent; wood-yards and gravel-banks where flat-boats are loading, tug-boats, small steamers plying to and fro, and all the signs of busy life accumulate with every mile.

As we approach Portland we observe its new, yet thrifty, appearance; the evidences of forests sacrificed to the growth of a town; and the increasing good taste and costliness of the buildings going up or recently built in the newest portions of the city. A low, level margin of ground, beautifully ornamented with majestic oaks, intervenes between us and the higher ground on which the town is built. Passing by this and the first few blocks of stores and warehouses, with their ugly rears toward the river, we haul up alongside a handsome, commodious wharf, and begin to look about us.

Portland is, we find, a cheerful-looking town of about 9,000 inhabitants; well paved, with handsome public buildings, and comfortable, home-like dwellings. It is at the head of ocean steam navigation, and owes its prominence as the commercial town of Oregon to that fact. Here the smaller steamers which ply on the Wallamet River have hitherto brought the produce of the valley to exchange for imported goods, or to be shipped on sailing vessels to foreign ports; and here has centered the commercial wealth and political influence of the State.

One hundred and ten miles from the sea is Portland, and twelve from the Columbia. At the first glance, this fact strikes the tourist with some surprise. But when he remembers that the shores of the Columbia are rough and heavily timbered, while the Wallamet Valley is an open, prairie country, his wonder vanishes. A town at this point was a commercial necessity, so long as the whole transportation business of the country depended on river communication. What effect to change commercial bases the opening of long lines of railroad will have, can hardly be determined before the drift of trade has defined itself. But, for the present, Portland is, in every sense, the chief town north of San Francisco.

From the relative importance of Portland to the other towns of the State, it deserves more than a passing notice. The site was first taken up, in 1843, by a man named Overton, from Tennessee. From him the title passed to Messrs. Lovejoy and Pettygrove about the beginning of the following year, during which the first dwelling—a log-house—was erected near the river, at the foot of what is now Washington Street. At this time the "claim" was covered with a dense forest of firs, which began to be cleared off, and the land surveyed into blocks and lots in 1845. A second building for a store was erected this winter, near the first one. It was not, like the dwelling, of logs, but a frame covered with shingles, and went by the name of the "Shingle Store" long after more ambitious competitors had arisen.

The growth of the embryo town was by no means rapid, as the year of its "taking up" witnessed the first considerable immigration to Oregon. Of these one thousand immigrants, a few stopped in Oregon City, the recognized capital of the Territory, and the remainder scattered over the fertile plains, in quest of the mile square of land for which they had come to this far-off country. The same continued to be true of the steadily increasing immigration of the following years; so that it was not until 1848 that Portland attained to the dignity of a name.

Of the two owners, one, Mr. Pettygrove, was from Maine, and desired the bantling to be called after the chief town of his native State. With the same laudable State love, Mr. Lovejoy, who was from Massachusetts, insisted on calling the town *Boston.* To end the dispute a penny was tossed up, and Mr. Pettygrove winning, the future city was christened Portland. When it is taken into consideration that Portland, Maine, is nearly two degrees farther south than Portland, Oregon, and that roses are blossoming in the gardens of the latter, while snow lies white and winter winds whistle over the leafless gardens of the former, the older city has no occasion to feel concerned for the comfort of its godchild.

After being named, Portland changed owners again. Mr. Pettygrove bought out his partner, and afterward sold the whole property to Mr. Daniel H. Lownsdale, receiving for it $5,000 in leather, tanned by Mr. Lownsdale in a tannery adjoining the town site. In 1848, or before the gold discoveries, *money* was almost unknown in Oregon ; orders on the Hudson's Bay Company, the Methodist Mission, and wheat, being the currency of the country. Mr. Lownsdale, it seems, had the honor of introducing a new circulating medium, which was Oregon-tanned leather.

Still another change in the proprietorship occurred in 1849—Lownsdale selling an interest in the town to W. W. Chapman and Stephen Coffin. During this year—there being now about one hundred inhabitants

—the Portlanders organized an association, and elected trustees for the purpose of erecting a building to be used as a meeting-house for religious services, and for a school-house.   It was used also as a court-room; and continued to serve the public in its triple capacity for several years.

The gold excitement of 1848–9 for a time had a tendency to check improvements in Oregon ; but finally the wandering gold-seekers began to return, and cultivate their neglected farms.   California demanded grain and lumber ; and these things Oregon could furnish in abundance.   Vessels now came frequently to Portland from San Francisco and the Sandwich Islands ; and in 1850 Couch & Co., of Portland, dispatched a vessel— the brig *Emma Preston*—to China.   In the same year Captain John H. Couch had his land claim surveyed into town lots, and formed what is known as "Couch's Addition," on the north side of town.   In this year, also, the pioneer steamboat of Oregon—the *Lot Whitcomb*—was launched on the Wallamet at Milwaukee, on Christmas day ; and the pioneer newspaper of Portland—the *Weekly Oregonian*—was started by Thomas J. Dyer.

In January, 1851, Portland was incorporated, having then about one thousand inhabitants.   In April the city officers were elected, and Hugh D. O'Bryant chosen Mayor.   Oregon having been erected into a Territory by the Act of Congress, 1848—her Governor arriving in the spring of '49—an election for delegate to Congress was held in June, 1851 (just twenty-one years ago), at which Portland cast two hundred and twenty-two votes.   In March of that year began the regular monthly mail service between Portland and San Francisco, per the steamship *Columbia*, Captain Dall.

The first brick building was erected in 1853, by Mr. William S. Ladd. Two years later the city boasted four churches, one academy, one public school, four printing-offices, about forty retail-stores of various kinds, one steam flouring, and four steam lumber-mills. The taxable property of that year was valued at $1,195,034, or about half the actual value of the real and personal property of the town.

From this time on the growth of Portland has been healthy and uniform. During the mining excitement of the upper country in 1864-5-6, there was a more hurried growth, and more inflated condition of trade, which subsided, however, with the excitement which occasioned it. Notwithstanding, there has been more costly and substantial improvement, both public and private, within the five years last past than ever before. Some of the business buildings and stores erected within that time are of truly metropolitan elegance and dimensions. The Court-house, the new Methodist Church, and the Custom House and Post-office, are large and costly edifices.

The city has fine public schools, and more than an equal number of seminaries, academies, and private schools. The Portland Academy, a Methodist institution, is a flourishing school for pupils of both sexes. St. Helen's Hall, a seminary for young ladies, is under control of the Bishop of the Episcopal Church; who also has recently established a grammar-school for boys. Both these seminaries are in a very flourishing condition. The Roman Catholic Church, also, has two schools, and the Jewish population one.

Of churches, Portland boasts a goodly number: two Roman Catholic, one Methodist, two Episcopalian, two Jewish, one Baptist, one Presbyterian, one Congregational, one Unitarian, one Lutheran, and one chapel for the colored population. Numerically the Catholic and Methodist denominations are the strongest, and the Unitarian the weakest.

Public amusements are only tolerably well supported. A fine theatre, however, is in course of construction, which will supply a public want. An Academy of Music, and a Musical Society, supported principally by the Jews, give occasional entertainments; and the brass bands are in the habit of discoursing sweet sounds upon the Plaza one or two afternoons in a week, when all the youth, beauty, and fashion of Portland come out for a promenade. A skating-rink furnishes amusement to the lovers of that exercise. Driving fast horses is quite a fashionable recreation; and an exhibition of Oregon stock is by no means an inferior one. A public library, comprising four or five thousand volumes, with a handsome chess-room in connection, offers attractions to the visitor and resident alike. The Young Men's Christian Association have also a reading-room in the same building.

The Odd Fellows have four associations, and a very fine temple; the Masonic Order, three associations and an elegant building; the Good Templars have three lodges, and there are several benevolent societies besides. The Fire Department consists of over two hundred active members, with two steam fire-engines, two hand-engines, and one hook-and-ladder truck, and hose carts. The department is very efficient, and large fires are of rare occurrence.

There are in Portland three large book-stores, and one or two stationery stores; three daily and weekly political newspapers, and one religious paper, published weekly; there are four banking-houses, including a branch of the Bank of British Columbia, besides half a dozen brokers, and several real-estate agents. All the ordinary branches of business are well represented, and the amount of taxable property in Portland is assessed at a value of between five and six millions. Its actual value is probably much greater. The city tax amounts to ten mills on the dollar. It is said that Portland is one of the richest towns of the size in the United States. There are ten of its business men whose incomes range from $16,000 to $50,000 ; ten more with incomes ranging from $8,000 to $12,000 ; and ten more having incomes of from $5,000 to $8,000 ; besides the capital owned and controlled by the Banking, Navigation, and Railroad companies. The improvements in the city for the year 1871 amount to $2,000,000.

Portland has a good drainage, the ground rising gently back from the river. It is at present supplied with plank sewers, which are generally kept in good condition. A water company supplies the city with water; and a gas company furnishes gas for lighting the streets, public buildings, and stores, and such private dwellings as are not too remote from the mains. The surveyed limits of the city include about three square miles ; the higher ground at the back being very desirable for residences from its superior healthfulness, and the fine views to be obtained. From any part of the city a quarter of a mile back, Mount Hood is seen in its finest aspect, rising grandly above the fir-clad slopes of the Cascade Range. It marks the place of the sun's rising in the summer months ; and passes at evening, when reflecting the hues of sunset, through many beautiful gradations of light and color. Even by moonlight its spectral shape is still discernible at

11

the distance of sixty miles. St. Helen, also, may be seen from the upper part of the city; and, from some points, Mount Jefferson.

Portland has not a "back country"—that is, it is divided from the agricultural portion of the valley on the west by the mountain ridge, which, commencing some miles south of this place, follows the west bank of the Wallamet to its lower mouth. The road which leads to the plains leaves Portland by a narrow ravine, and, following the pass of a stream, crosses the mountains through a dense forest of firs and pines. It is a pleasant-enough drive in summer, but quite the reverse during the rainy season. By the beginning of 1872, however, the Oregon Central Railroad will have been completed from this city to the town of Hillsboro, a distance of twenty miles—four or five miles beyond the timbered ridge. There is a beautiful, short drive of macadamized road, extending about six miles south of town, along the bank of the river, and terminating at the Milwaukee Ferry, or the "White House." The road down the river is not a good one, though a very little expense would make it so, and it might be continued all the way to St. Helen, making a pleasant and useful highway; but the small steamers that run on all the rivers have made roads of secondary importance near the margins of these streams.

The river in front of Portland is about one-quarter of a mile wide, with water enough for large vessels to lie in; and the rise and fall of the tide amounts to a couple of feet. During the winter flood in the Wallamet, which is occasioned by heavy rains, the water rises about eight feet. For this reason the wharves are all built in two stories—one for low, and one for high-water. The great flood of 1862, and that of 1870, brought the water over the upper wharves and even over Front street, which is twenty-five feet above low-water mark. The summer flood in the Columbia, occasioned by the melting of snow in the mountains where it has its sources, backs the water up in the Wallamet as far as the falls at Oregon City, which again makes it necessary to abandon the lower wharves. These two rises keep this portion of the Wallamet supplied with water through the greater portion of the year; but it is necessary to dredge the channel below the city in the latter part of the summer. Since the dredger came into use no vessels have been stopped by bars, but all discharge their freight at the wharves.

There is a regular line of ocean steamers belonging to the North Pacific Transportation Company—Holladay and Brenham, owners—which makes three or four trips a month between San Francisco and Portland; and another line owned by the same company, making about the same number of trips to Victoria and Puget Sound. The Oregon Steam Navigation Company run steamers upon the Columbia River and several of its smaller tributaries—semi-weekly to Astoria, daily to the Cowlitz River and intermediate points, and daily to the Cascades and Dalles City; semi-weekly, or tri-weekly, from Dalles to Wallula; and at stated periods on the Snake River and Northern branch of the Columbia. Sailing vessels run quite regularly to the Sandwich Islands, China, South America, and New York, as well as to San Francisco; and the trade is yearly increasing.

The facilities for travel in the Wallamet Valley, and southward, are: first, the Oregon and California Railroad, which is already in running order to the head of the valley on the east side, connecting with a line of

stages to Red Bluff, in California; and, secondly, a
steamer line on the Wallamet River, owned by the
Railroad Company, and plying between Portland and
all points on the river north of Eugene City, when there
is a sufficient stage of water. A stage line starting
from Portland on the west side of the valley, carries
passengers to Corvallis, where they connect with the
railroad. The Oregon Central, or "West Side" road,
will soon do away with staging through this portion of
the valley also. Travel by land is by no means difficult
in any portion of the Wallamet Valley, the roads being
excellent and the conveyances good. On some of the
smaller streams there are steamers plying, connecting
with the main lines of travel; and each year increases
these facilities for locomotion.

Portland is well supplied with hotels, which in general answer very well to the awkward guest's description of his dinner, "Good enough, what there was of
it; and enough of it unless it were better." The latest
built, and very well conducted, is the St. Charles, on
Front Street. Another and larger one will soon be
finished near the steamer landing; but the town seems
to need a commodious hotel farther back from the
river, away from the confusion and crowd of business
movements.

East Portland, on the opposite side of the river,
contains about one thousand inhabitants. It has a
fine, level site, and a pleasant country back of it.
Considerable importance attaches to it on account of
its being the initial point of the Oregon and California
Railroad, and the location of its machine-shops and
warehouses. A steam-ferry connects it at present
with Portland, and it is soon to be united to the latter place by a bridge over the Wallamet, the contract

for getting out the timbers having been given to the St. Helen Milling Company. Improvements are rapidly increasing in East Portland, and property is held at a pretty high figure. The Railroad Wharf is 1,250 feet long by 70 in breadth; is built in a slope to accommodate it to the different stages of water, and has a substantial warehouse upon it 370 feet long by 42 wide—being just half the size of the corresponding wharf and warehouse on the west side of the river.

East Portland contains some fine residences, several churches, and a bank, and supports a newspaper of its own, as well as several societies and orders.

About two miles above the town are the machine-shops and car-building establishments of the Oregon and California Railroad; and about four miles above town is the Company's saw-mill—one of the largest and most complete in the State. When in full operation it employs sixty men as sawyers, attendants, loggers, and drivers; and is capable of cutting 1,000,000 feet per month.

CHAPTER XIII.

THAT portion of the Wallamet between Portland and Oregon City, a distance of twelve miles, is very charming, in a quiet, picturesque style. On the east side the country is level, the banks being moderately high, and well wooded. On the west side the mountains keep along, at a little distance back from the river, for some miles. We pass by a skirting of bottom - land, with a belt of oak - trees on a slight ridge, and soon come abreast of Oak Island, a longish, narrow island —covered with a growth of fine, large oak - trees—on which is a house or two. The island, we believe, is used as a milk - ranch, the alluvial nature of the soil making it a good piece of pasture - ground.

A romantic bit of history is told in connection with this island. In the early times of American occupancy in Oregon—that is, about 1841—a half - dozen young men who had strayed to this remote corner of the world, where they found nobody except fur - traders and missionaries, became dissatisfied with a country where there were no white women whom they could marry; and being determined not to take Indian wives, as too many others were willing to do, resorted to this island, and together built a small schooner, with the purpose of getting to California. The only place in the country where they could procure sails, cordage, and rigging generally, was at the Hudson's Bay Company's post, Fort Vancouver. But Dr. McLaughlin looked upon their scheme as a hair-brained one, and refused to assist them to peril their lives in the manner proposed.

However, the United States Exploring Squadron happened upon the coast about that time, and the young men complained to Lieut. Wilkes that Dr. McLaughlin would not aid them, giving what they supposed to be the reason. Upon Wilkes representing the wishes and firm determination of the adventurers to the Doctor, he consented that they should be permitted to sacrifice themselves in their own fashion. Their vessel was supplied with every thing requisite, they went to sea in her, reached San Francisco in safety, and sold their little craft for a handsome sum—actually making a very good speculation out of their seemingly desperate undertaking. It is to be hoped that when they reached home, they found young women deserving of such heroic perseverance and unflinching bravery. The name of their lucky little craft was, *The Star of Oregon*.

Above Oak Island the river narrows somewhat, but preserves its attractiveness. The first settlement aspiring to be reckoned a town is at Oswego, about six miles up, on the right bank ; and is celebrated for being the first and only iron-smelting establishment in the State. The smelting-works were erected in 1867, at a cost of $100,000 ; but owing to some necessity for paying a heavy royalty for the privilege of taking out the ore, was not at first considered profitable. Nevertheless, considerable pig-iron, of the best quality, has been manufactured here, both for Portland and San Francisco foundries. There is also a large lumber-mill at this place.

At Milwaukie, on the east or left bank of the river, is the famous "Standard" flouring-mill, which exports "best Oregon," in large shipments, to San Francisco, the Sandwich Islands, and New York. Here, too, is the first nursery of the Pacific Coast. From the grounds of Meek and Lluelling, at this place, were taken the first cultivated apple-trees ; and the earliest export of this fruit was made to San Francisco in 1853, when two hundred pounds brought five hundred dollars. The following year the same firm sold forty bushels for $62.50 a bushel ! From that time to this Oregon has maintained its reputation for apple-raising, until "the land of red apples," or "the land of cider," has come to be its synonym.

Milwaukie is one of those towns that started in as the rival of some other town—Oregon City, in this instance—and could not sustain itself. It has, however, great advantages for milling and manufacturing, from the abundance of water-power in its vicinity available for these purposes. The Wallamet receives within a dozen miles three tributaries, either of which is a good milling stream. Milwaukie has a large tannery, which turns out as good leather as is made on the coast—a branch of business very profitable in this country.

As we approach Oregon City the river becomes quite narrow in places, and in summer, when the water is low, the channel is barely wide enough for the steamer to pass between the gravel bars. The attention of the tourist is first attracted, on nearing the town, to the spray, which rises like a mist from the river, just above the steamer's landing, and he gazes with ever-increasing interest upon the leaping, foaming cataract of the Wallamet, which, although less in height and in volume than Niagara, has much of the same grandeur and force.

Formerly it was necessary to make a portage of more than a mile around the falls ; but the basin, constructed at a great expense by the People's Transportation Company, now enables boats to come down to the warehouse, and the passengers are transferred by simply passing through a long, covered building to the boat lying in the basin at its upper extremity. From the deck of the second steamer a perfect view of the falls is obtained.

Oregon City stands upon a bed of basaltic rock—a ledge of which extends quite across the river, and crops out on the opposite side. This ledge is about twenty feet higher than the surface of the water below the falls, and worn and broken into a jagged crescent, with rather a sharp angle in the centre, where the river deflects toward the western shore. In low or ordinary stage of water the stream divides into several parts, seeking the deepest channels in the rocks, and forming a number of different cataracts ; yet the central one, at the angle spoken of, is always the principal one. Above the falls the river parts, flowing around an island of rock, on which once stood a mill belonging to the Methodist Mission, but which was carried away in the great flood of 1862, along with numerous other buildings from the mainland.

The current, always strong just above the falls, is terrific when the heavy rains of winter have swollen all the tributaries of the river, and filled its banks with a rushing torrent fifteen to twenty feet in depth. At such times the rocks are mostly hidden, and the falls extend from shore to shore, or about a quarter of a mile. In the early history of the country, a party of four persons—two gentlemen and two ladies—with their two Indian boatmen, were carried over the falls by the force of the current while attempting to make a landing above. A few years later a small steamer became unmanageable, and was borne swiftly to destruction in the maelstrom below the central fall, carrying with it the captain and three others to an appalling death. The current which sets into the "basin" in high-water is alarming to the nervous passenger; and a steamboat is often an hour in getting out of it into the river above, during which hour he has plenty of time to imagine all that might happen should the machinery become disabled, or the cables part, which, for greater safety, connect the boat with the shore. In ordinary stages of water there is no difficulty in contending with the force of the water. A canal around the falls is in course of construction, which will do away with the portage entirely.

The Falls of the Wallamet constitute the great water-power of the State. The favorite term for Oregon City is, "The Lowell of the Pacific Coast;" and there is indeed every natural agency here for the making of a second Lowell. One of the largest woolen-mills of the State is located here. It is built substantially of stone and brick, four stories high, and 190 by 60 feet in ground area; and contains twelve sets of the most improved machinery. Its manufactures are blankets, flannels, and cassimeres, and light cloths. It is the intention of the Company in time to manufacture delaines, and other fabrics used for women's dresses.

The celebrated "Imperial" flouring-mill is located at Oregon City, which has a capacity for turning out five hundred barrels of flour every twenty-four hours. Another flouring-mill; a paper-mill, for the manufacture of coarse and printing paper; a lumber-mill, machine-shops, and other industries, show the business resources of the place, which, although the oldest town in Oregon, is yet only a small one.

Oregon City, like Portland, has a good number of churches — Methodist, Episcopal, Catholic, Baptist, and Congregational. A seminary, and graded public school, besides two or three private schools, furnish educational facilities. A Government Land Office is located here, where the lands for the northern portion of the State are entered. The necessary transfer and handling of all freights intended for the valley, or coming from it, gave Oregon City formerly a great deal of business. The opening of the Oregon and California Railroad may divert a portion of this freight from the river, but there must always remain a much larger amount of the transportation of the valley which will seek the cheaper water-carriage.

Oregon City was first claimed by Dr. John McLaughlin, of the Hudson's Bay Company, in 1829, who commenced to build a saw-mill at the falls. Three log-houses were erected that winter, and timbers squared for the mill; but the building progressed no further at that time. Some portion of the land was planted to potatoes in the following spring; and in 1832 the mill-race was blasted. The houses built in 1829 were destroyed by the Indians, and replaced in 1838 by one small dwelling and store-house; and the square timbers for the mill were hauled upon the ground, but not put together that year. In the meantime the Methodist Mission asked, and obtained, permission to build upon the claim, which they did in 1840, erecting a dwelling and store-house in one; and Mr. A. F. Waller took possession of, and resided in, this building.

Disputes then arose as to the possession of the

claim, and a series of aggressions, concessions, and compromises took place. In the meantime the Mission opened a store, receiving a fresh supply of goods in 1842. There was also a Milling Company formed this year, which proceeded to build a saw-mill on the island, already mentioned. Several gentlemen came out from the States that fall, one or more of whom entered into trade at Oregon City, or Wallamet Falls, as it was then called. In the following year quite a large immigration arrived; such persons as did not desire to reside upon farms, congregating at this place. Soon a Provisional Government was talked of, was finally decided upon, and adopted. Oregon City became the recognized capital, as it was the principal seat of American enterprise in the Territory. As early as 1846 it boasted a newspaper—the *Oregon Spectator*—a seminary, and a debating club. Its pretensions to literary attainments, judging from the contributions to the *Spectator*, were very well founded. From the same source we learn that it was not without its social entertainments, its local politics, and other excitements—and, from the advertising columns, that almost every branch of business common to the ordinary town life was even then represented; while an export trade was carried on by Dr. McLaughlin, and in a measure by the Americans—the articles exported being lumber and wheat.

Oregon City continued to maintain its supremacy until the gold discovery in California, which, giving rise to an active commerce with that country, imparted to Portland an impetus that soon enabled it to outstrip the city at the falls, which had not the required depth of water for floating an extensive merchant navy. Railroads will ultimately remove any disabilities of that kind, and with its splendid water-power, backed by a country productive in soil, timber, and mineral deposits, its future seems as well assured as that of any town in Oregon.

Canemah, a mile or more above Oregon City, and ultimately to be a portion of it, was the upper landing when the People's Transportation Company used to transport their freight and passengers around the falls by a horse-railroad. It is rather a more favorable site for building than just about the falls, where a high bench of trap-formation crowds the lower portion of the town quite to the river-bank.

Opposite to Oregon City is another of those abortive attempts at a town for which this country is rather remarkable. Of this one, nothing now remains but one or two decayed buildings, and the name—Linn— after that Missouri Senator who introduced the Oregon Land Bill, of 1843, which occasioned the immigration of that year.

From Oregon City, for a distance of more than fifty miles by the river, there are no towns of any importance; though there are numerous "landings," where freight is put on or off for various places in the interior, indicating that there is a considerable population scattered through the valley. The scenery of the Wallamet is of rather a monotonous character, though pretty—the best portion of it being between Portland and Rock Island, above Oregon City a short distance. After this is passed, we begin to wish away the belt of timber which hides the level country back of it. It is not, however, until about twenty or twenty-five miles have been passed above Portland, that the prairie country commences; the lower portion of the Wallamet Valley toward the Columbia being heavily timbered.   Even when we have come opposite to the open plains, there is still a screen of trees between us and them.

It is apparent that most of the level country lies on the eastern side of the river, and that a chain of hills crosses the west side of the valley transversely. Some of the high, rolling land of the west side offers beautiful farm-sites, preferable for their splendid views and sunny exposures to the level plains. Fruit, it is understood, does better upon these farms than upon those of the prairie.

Steaming along up at rather a low rate of speed, there is little to entertain the traveler, except the frequent windings, the luxuriant vegetation of the river-banks, and observations on the current, which is often very rapid.   There is a fall of 400 feet in the 125 miles between Portland and Eugene City, at the head of high-water navigation on the Wallamet.   This being true, rapids might reasonably be looked for in this river.   At no place, except at Oregon City, is navigation seriously impeded by them; but very frequently they give the shallow, narrow hull of our boat all it can do to make its way against them.   The water is beautifully clear, and the bed of the stream has a gravelly bottom.

Mention should be made of Champoeg, the French-Canadian settlement of the retired servants of the Hudson's Bay Company, on the east side of the river, twelve or fifteen miles above Oregon City.   It was here that the "Organic Laws" were adopted by a majority of the Oregon settlers, in May, 1843, and a Provisional Government erected, to last until such time as the United States Government should see fit to acknowledge Oregon as one of her Territories.

About twenty-five miles above Oregon City, the Yamhill River enters the Wallamet from the west. It is a narrow stream, and its entrance is almost hidden by the profusion of overhanging shrubbery and trees. Waiting here for freight and passengers, is the *Dayton*, a commodious steamer of light draught, which will convey us, if we so elect, to the towns of Dayton, La Fayette, or McMinnville, in Yamhill County — one of the finest agricultural portions of the State, and celebrated for having domiciled, at one time or another, almost every person of prominence in the State, prior to 1868.

Above the Yamhill, the traveler sees nothing of interest, beyond a wood-yard or a grist-mill, all the way to Salem. There is, however, a memorable spot twelve miles below Salem, on the east bank, where the Methodist Mission made its first location in 1834; this being the very first American settlement in the Wallamet Valley. Here, too, in 1843, after the acceptance of the Organic Laws, was held the first Legislative Assembly of nine persons; their Council Chamber being a public room in a building belonging to the mission, known as "The Granary." Subsequently, the Legislature removed its sessions to Oregon City. The high-water of 1862 carried away a portion of the old mission ground, which was situated on the bank of the river, where the open prairie approaches quite to it.

CHAPTER XIV.

WHILE we are overcoming the last twelve miles of quiet voyaging between the "Old Mission" and Salem, we may as well consider their relationship. In the autumn of 1840, the Methodist Mission built a mill on a stream twelve miles south of their first establishment, at a place called by the Indians Chemeketa, and finding the situation every way a better one than that, removed the mission to it in the following year. The first dwelling was erected at some distance back from the river, on the bank of a stream known as Mill Creek, in a very pleasant and convenient location, with an extensive plain on one hand, and a charmingly wooded, rolling landscape on the other. In 1843, the large frame-building, for many years known as "The Institute," was erected, as a school for Indian children; but the savages not taking very kindly to study, the mission was dissolved in 1844, after which time the Oregon Institute became a seminary of learning for whoever chose to patronize it, although it still remained under the control of the Methodist denomination, and was converted ultimately into a university.

Upon the sale of the mission property, the town-site of Salem was laid out by Mr. W. H. Wilson, and received its present name. It is very handsomely located upon a gravelly prairie, rising gradually back from the river, which is skirted with groves of tall trees. Other groves of firs and oaks relieve the level monotony of the landscape for a couple of miles away to the north and east; while the hills across Mill Creek are wooded like parks, with a variety of trees. Across the Wallamet, and fronting the town, is a range of high land called the "Polk County Hills," which makes the greatest charm of the whole view of Salem. In outline and coloring, these hills are poetically beautiful. Should we chance to drive in the direction of the Penitentiary grounds (east), a lovely landscape lies stretched on either side, melting and blending into one complete picture. The town is backed by the Polk County Hills, to the west; the "Waldo Hills" (another arable range), to the south-east; the blue Cascade Range with its overtopping snow-peaks, to the north-east; groves of fine, large oaks and firs breaking the middle distance; while immediately about us are level farms and fields of waving grain, with a substantial farm-house, here and there, in their midst.

Salem is a comfortably built town, with an air of stability and propriety about it. The streets are wide, the lots large, and the dwellings neat, with well-kept gardens attached. Shade-trees—locust and maple—line the broad avenues; and the public square is of liberal proportions, promising "lungs" to the city, should it grow large enough to need this breathing-space in its midst. Although the capital of Oregon, it has as yet no State buildings upon its spacious square. During Territorial days, there was in Oregon the usual struggle between rival towns to secure the capital. Salem, having triumphantly carried off the honor and the Government appropriations, had bitter enemies, as might be expected; and when the handsome State House was near its completion, it was unfortunately and mysteriously destroyed by fire. Since then, the State has rented apartments in a brick block on the principal business street, where the public archives are kept, together with the State Library, and where the Legislature holds its biennial sessions.

Notwithstanding this lack, the town is not without some of the handsomest buildings in the State. Reed's Opera House, the Chemeketa Hotel, the Bank building, the new Wallamet University building, and some of the stores, are quite worthy of an older and wealthier city. The private residences, too, are many of them spacious, and even elegant. Taking it altogether, Salem is probably the pleasantest town in Oregon; and from its central location, together with its importance as the capital, can never be less than the second city of the State. It has now connection with Portland by the Oregon Central Railroad; and very soon will be, by the West Side Railroad, connected with the country bordering on the Columbia River.

The Agricultural Society of Oregon have their Fair Grounds at Salem, where annually are congregated the rural population from every part of the State. Those who come from a distance are provided with tents, beds, and cooking utensils; the fields adjoining the inclosed grounds swarming with these families, their tents, wagons, and animals. The occasion is employed to renew old acquaintances, and talk over the politics and agricultural interests of the country. Each year witnesses some improvement in stock or machinery; and the articles on exhibition are very creditable, for a State with so limited a population. The prizes offered are liberal, when the resources of the Society are considered.

The manufactures of Salem are: one woolen-mill,

valued at $300,000, turning out yearly $200,000 worth of blankets, flannels, tweeds, cassimeres, yarn, and knit goods; two flouring-mills, both making an excellent brand of flour; one oil-mill; two tanneries; three lumber-mills; four sash and door factories; one foundry and machine-shop; four wagon and carriage-shops; two cabinet-shops; one bag-factory; three printing-offices; one book-bindery; two gun-shops; three breweries; three saddle and harness manufactories; and four millinery establishments. There is one banking-house, numerous dry-goods and grocery stores, three bookstores, three drug-stores, and two hardware stores. Not that every branch of business in Salem is comprised in this list; but this is a general summing up of the industries of a population of about four thousand people, outside the professions and the agricultural classes. Probably the assessable property of Salem amounts to two millions or more.

Salem has one daily and weekly newspaper, the *Unionist*, and *Statesman;* and one other political weekly, the *Salem Mercury*. The *Wallamet Farmer*, an agricultural journal, is also a weekly, having its publishing office in Salem; and the *Oregon Medical and Surgical Reporter*, a monthly, is also published here. The educational facilities of Salem are good. The Wallamet University, with a Medical Department, takes the first rank. The old Institute building having become somewhat dilapidated, the present structure was erected in 1864. It is built of brick, in the form of a Greek cross; is eighty-four feet in length by forty in width, and is five stories high, including the basement and attic. The plan of the interior is convenient and elegant. There are three entrances, and three separate staircases leading to the attic. From the cupola a splendid view of the country is obtained, with four snowy peaks in the distance. The university is in a good condition financially, and ranks among the ablest institutions of learning on the Northern Coast.

The Catholics have a seminary in a flourishing condition, and there are fine public-school buildings for the accommodation of the public generally; but the free-school system is not yet put in operation in any part of Oregon. Salem has nine churches, comprising all the usual denominations; a musical society; three lodges of Odd Fellows, and one of Masons, and a Good Templars organization.

The State Penitentiary is located east of town on a tract of seventy-five acres, belonging to Government, where also the Insane Asylum is ultimately to be built. None of the State buildings yet erected are of a permanent character. The contracts for these structures will be a "bone to pick" between rival contractors at some future day, and will give Salem a chance to make something out of them.

The value of property has considerably increased since the opening of the Oregon and California Railroad, and must continue to increase for an indefinite period, as the growth of Salem is assured. The gradual settlement of the choice farming lands by which it is surrounded, and the opening up of the mineral deposits known to exist in the Cascade Mountains to the east, would alone give a sure, if gradual, rise to property in Salem. Its population is order-loving, social, and industrious; and its climate healthful. Intermittents prevail to some extent on the margin of the river, as in all countries, especially new ones; but they are of a light character, and easily broken up, or what is better, prevented.

It is a fact, more or less well established by experience, that the sunflower is anti-miasmatic. It is asserted, on very good authority, that if a hedge of this plant be interposed between the dwelling and the prevailing direction of the wind, or if the dwelling be surrounded by them, that intermittent fevers will not attack the occupants. As the seeds are useful for some purposes, it could be no loss to any one residing in fever-breeding localities to try the effect of cultivating them.

South Salem, a pleasant suburban neighborhood, separated from the city proper only by Mill Creek, is rapidly becoming an important addition to it. Many of the pleasantest homes are located in this neighborhood, which, from being rather more elevated than Salem, is in demand for the beauty of its building sites, and the extent of its river views.

Driving anywhere about Salem is delightful. The roads are naturally good, except in the rainy season. In summer the town-people enjoy excursions to the mountains, for trout-fishing, strawberrying, and the like amusements. It is by no means uncommon for parties to camp out for one or two weeks, either in the mountains or over on the sea-coast. The summer climate being generally rainless, there is no risk in this nomadic sort of life; and people find themselves the better for this intimacy with Nature. Another summer custom is the holding of "basket-meetings," for a week or more, by several of the churches, who have grounds set apart, and the necessary buildings thereon, for these annual gatherings.

The Salem people have two of these camp-grounds, adjoining each other, over the river in Polk County, on the banks of the Rickreal, near its junction with the Wallamet. Here, in a fine grove of firs, we have seen the order and devotion, the sociality and recreation, of a basket-meeting. Between the hours of service the people disperse themselves in all directions, to lunch, and to talk over church affairs—perhaps the prospect of a crop; for this is the season of rest for the agricultural population—between "seed-time and harvest."

The scene is very picturesque. White tents, and rough board cabins, are thickly placed among the trees. In the centre of the grove is the spreading roof, supported on rustic pillars, under which the congregation gathers at stated hours for religious services, and where the speaker's desk is placed, with it great bouquets of roses and sweet-scented garden flowers—contributions from the ladies to the adornment of the rude pulpit. Here and there a covered wagon serves as a temporary home; for many of these people crossed the plains years ago, and know to how many uses a covered wagon may be put. Young people are flitting about from tent to tent—older ones are receiving company at their own doors; tables are spread in the shade, at which hungry people do justice to hasty cookery; a hum of subdued voices fills the air with a pleasant murmur, which accords well with the soft sighing of the trees, the stir of insects in the air, and the flow of the pebbly stream close by.

That "the groves were God's first temples" strikes us forcibly under circumstances like these. The devotional spirit comes more easily and quickly, and with more power, in immediate contact with Nature, than when coaxed and stimulated into exercise by the appliances of art. In the age when architecture was really and truly an art, this truth was seized upon;

and those grand cathedrals which still remain the glory of Europe, in their pointed roofs, fretted arches, and long colonnades; their deep shadows, and windows of colored glass, staining the light they transmitted to the colors of Nature's choicest hues, were intended to express that solemn and subtile sense of beauty, which, in the presence of great Nature, lifts the heart above and away from mean or trivial considerations.

The Salemites have some other resorts than those already mentioned, in different soda-springs, in their own, and the adjoining county of Linn. In short, if the tourist has not remained in the heart of the Wallamet Valley long enough to find out for himself its resources for pleasure, as well as profit, he has done himself and the country an injustice. Of course all the resorts mentioned are frequented by residents of the adjoining counties on either side, and belong equally to all this portion of the State. It is, indeed, quite the custom for Oregonians, of every section, to make their summer excursions, quite as much as those city-bred pleasure-seekers who people Eastern watering-places every season.

About twenty miles above Salem the Wallamet receives the waters of the Santiam, a considerable stream, having its rise in the snows of Mount Jefferson. Lebanon, on the south fork of the Santiam, is a delightful spot, in the midst of a fine farming country. A few miles above Lebanon, at the falls of the Santiam, is another small town, with flouring and lumber-mills. Both of these places are the centres of a healthy business, dependent on agriculture and manufactures.

CHAPTER XV

ALBANY, AND OTHER RIVER TOWNS.

ALBANY, on the east side, is twenty-five miles south of Salem, in a tolerably straight line ; by the windings of the river it is farther. It is about the head of river navigation in the low water of late summer. Between Salem and Albany are several small places, of no particular importance, chiefly on the west side of the river. At one of these—Buena Vista—considerable coarse earthenware is manufactured. Monmouth, near the river, is the seat of Monmouth College, under the control of the *Christian* denomination. Warehouses and shipping points are frequent along this portion of the river; for some of the most famous grain-raising counties border it. The Oregon and California Railroad connects this town with those already mentioned, and has already added a considerable interest to business, and value to real estate.

The Calapooia River enters the Wallamet at Albany. This stream furnishes fine water-power up in the foot-hills, where two towns—North and South Browns-ville—are located. The former is a manufacturing place, having a woolen-mill, a flouring-mill, a planing-mill, and a tannery, besides machine-shops, and other similar establishments.

Albany was laid out as a town-site in 1848 by two brothers, Thomas and Walter Monteith. All that has been said of Salem, as a well-located town, applies equally to Albany. It is hardly less beautiful, none the less industrious, thriving, or intelligent ; and is the third town in importance in Oregon. With a population of twenty hundred, it has four churches ; a college building ; the best court - house, out of Portland, in the State ; a fine public-school building ; two flouring-mills ; two lumber-mills, and good, substantial brick stores in proportion. Every trade and industry is well represented ; and the character of its people is not below that of any town of its size on the coast ; while its business men are noted for their enterprise and public spirit. We are pleased to pay this tribute to Albany, where we met some very congenial people. There is no place in the interior of Oregon where the stranger is more likely to be pleased with his surroundings than here.

There are also many pleasant drives and resorts about Albany, and a fine view of that beautiful group of snow - peaks, the Three Sisters. Although there is much level prairie, there are also buttes and ridges so disposed about the valley as to give a charming variety to an otherwise monotonous landscape. Opposite Albany, on the west side of the river, is a belt of heavily timbered bottom - land, which is subject to overflow, and back of that rise the rolling hills of Benton County, dotted with magnificent spreading oaks.

Above Albany the pine-tree begins to appear, mixed with the fir, along the river - banks. The groves of timber are more scattering, and the country more level and open. Except the ash, maple, alder, and willow of the river - bottoms, there is little forest ; but the isolated trees of pine, fir, and oak which beautify the plains, are of the handsomest proportions.

Corvallis, about a dozen miles above Albany, on the

west side of the river, is about the same age with it. Its first proprietor was Mr. Avery, who still resides there. It is the first town of consequence on the west side of the Wallamet, and the only one excepting Eugene. The situation of Corvallis is remarkably handsome, having the river on one side of it, and the Coast Range sufficiently near it on the other to give the landscape the look of being framed in a semicircle of hills.

A road through the Coast Range directly west of Corvallis, furnishes this place communication with Yaquina Bay on the coast, thus giving it an independent sea-port. Besides this advantage which it affords to shippers, the bay has become quite a famous summer resort, through the facilities furnished by this road. The climate of Corvallis is also perceptibly affected in summer by the sea-breezes which find their way into the valley through the pass in the mountains along which this road conducts us. St. Mary's Mountain is a peak of the Coast Range in full view from Corvallis, and another summer resort for pleasure-seekers. One of the attractions is the delicious cream to be obtained from a dairy up on the mountain—which, with strawberries or huckleberries, is said to make a very fine dessert to a "basket" dinner.

Corvallis narrowly escaped being made the capital of Oregon Territory, and received instead thereof the appropriation for a State University. But the money was expended, and the only result is a pile of ruins— another example of how the Territory used the appropriations of Congress. The State is a much better economist.

Corvallis, with a population of ten or twelve hundred, supports three churches, an academy, and female seminary, besides common schools, and a college within a few miles of it. It has considerable trade ; though, having been cut off from river navigation fully half the year, it could not have a constant trade which was not purely local. As it is situated in one of the best agricultural sections, the time when the railroad reaches it, which will be very soon, will see a rapid change in that respect. The *Corvallis Gazette*, a weekly newspaper, is a well-conducted journal.

Two or three miles south of Corvallis, on the east side of the river, is a new town called Halsy, an outgrowth of the Oregon and California Railroad. It is receiving a considerable number of settlers, and promises to be a place of some importance as a grain-depot.

The face of the country in this portion of the Wallamet Valley is extremely picturesque and beautiful. The narrowing of the valley toward its head brings mountains, plains, and groves within the sweep of unassisted vision, and the whole resembles a grand picture. We have not here the heavy forests of the Columbia River region, nor even the frequently recurring fir-groves of the Middle Wallamet. The foot-hills of the mountains approach within a few miles on either side, but those nearest the valley are rounded, grassy knolls, over which are scattered groups of firs, pines, or oaks, while the river-bottom is bordered with tall cottonwoods, and studded rather closely with pines of a lofty height and noble form.

Two tributaries enter the Wallamet between Corvallis and Eugene—the Muddy, from the east, and Long Tom from the south-west. The country on the Long Tom is celebrated for its fertility, and for the uncompromising democracy of its people. The school-master and the Black Republican, are reported to be alike objects of aversion in that famous district.   It is also claimed for Long Tom, that it originated the term "Webfoot," which is so universally applied to Oregonians by their California neighbors.   The story runs as follows: A young couple from Missouri settled upon a land-claim on the banks of this river, and in due course of time a son and heir was born to them.   A California "commercial traveler" chancing to stop with the happy parents overnight, made some joking remarks upon the subject, warning them not to let the baby get drowned in the rather unusually extensive mud-puddle by which the premises were disfigured, when the father replied that they had looked out for that; and, uncovering the baby's feet, astonished the joker by showing him that they were *webbed*.   The *sobriquet* of Webfoot having thus been attached to Oregon-born babies, has continued to be a favorite appellative ever since.

No inland town could have a prettier location than Eugene, and few a more desirable one for other reasons.   At the head of the Wallamet Valley, it combines many advantages; Lane County, of which it is the county-seat, extending from the sea-coast to the Cascade Range, and including grain and stock lands, timber and mineral lands, with abundant water-power.   It is also the starting-point of the Military Road, crossing the Cascades at Diamond Peak Pass, and traversing Eastern Oregon near its southern boundary, to Owyhee, in Idaho.   It is presumable, at least, that this must be the course of a railroad at no very distant day.

Like all the towns in the Wallamet Valley, Eugene has recognized the value of the church and the school-house in the community.   With a population of about

nine hundred, it has five churches, an academy, and
other public and private schools.   In its early days,
for it was founded more than twenty years ago, an
attempt was made to have a college located here.   The
enterprise proceeded as far as the partial erection of a
handsome stone building, when it was arrested, and has
so remained ever since.   Before the completion of the
railroad the trade of Eugene could not be very great,
owing to the want of means for transporting the
products of the country to any other market than its
own.   Its inhabitants, however, enjoyed peace and
plenty in their own homes; and perhaps were more
intellectual and more social from their isolation.   The
literary professions are well represented, and the
trades seem to thrive as well as in more bustling
places.   The office of the Surveyor General of the
State is located here.

Three miles above Eugene is the new town of Spring-
field, already a thriving little place, with flouring and
saw - mills, and several manufactories.   Following up
the Middle Fork of the Wallamet, leads us through a
valley, heading in the Cascade Range, to the south-
east.   This valley, together with several smaller lateral ones, contains a considerable amount of excellent
land, both for grain - growing and stock - raising.   For
dairy purposes, much of it is excellent; also, for wool-
growing.   Fine water - power may be obtained in numerous places, owing to the rapid fall of the streams
coming out of the mountains.   It is up this valley
that the Military Road leads to the Diamond Peak
Pass.

It is claimed that up among these foot - hills every
variety of fruit and vegetables can be more successfully cultivated than on the prairie land of the great

valley.   Certainly it is evident that the resources of
this part of the country, in soil, timber, water, and
minerals, are unexcelled by any portion of it ; and
only its remoteness has hitherto prevented its settlement.   Already the lands are · beginning to be taken
up, and settlers' cabins to appear on frequent claims on
the Middle Fork of the Wallamet.   McKenzie, or
North Fork, is a large stream, with a similar country
and advantages for locating.   The South Fork is
smaller, with the same general character.

A glance at the map of Oregon will show any one
the horse - shoe shape of the head of the Wallamet
Valley, with the Coast Range on the west, the Cascades on the east, and the Calapooias on the south.

This amphitheatre of mountains, running down into
the valley in long slopes and ridges, furnishes it with
superior facilities for a great variety of manufactures
which depend on wood, water, stone, and such like
materials.   When these are to be found, together with
a variety of good soils, adapted to all branches of
farming, there can be no doubt of the future of such a
country.   From every side, the riches of these hills
will glide down into the lap of that city which is situated in their midst.

CHAPTER XVI.

THE prairies of the Wallamet Valley are not an uninterrupted level, like those of Illinois. In some parts they resemble the "oak openings" of Michigan; again they are quite extensive plains, but nowhere out of sight of large bodies of timber, either on the mountains or along the Wallamet, and its numerous affluents. Ranges of hills and isolated buttes occur frequently enough to save the landscape from monotony, and to furnish variety in soil and location.

Time was when all this valley waved in early summer with luxuriant native grasses, red and white clover, and beautiful wild flowers. When the first herds of California cattle, purchased in that country, and driven over the mountains of Southern Oregon, with great labor, and danger from the hostility of the Indians, to supply the Mission and the earliest settlers, they might wallow through grass breast-high on the prairies, and higher than their heads in the creek-bottoms. These herds increased rapidly; and the country being sparsely settled, they were allowed to roam at will over it.

Stock-raising was an easy and lucrative business in Oregon at an early day: in the first place, because cattle were scarce among the settlers; and next, because after they had become more numerous, they came suddenly into demand as food for the freshly imported mining population with which the gold discovery flooded

the southern portion of the State. The stock-owner
put his brand on his herd, and turned them out to
"summer and winter" themselves on the abundance of
the virgin prairies. In course of time this indiscrimi-
nate pasturing injured the grasses, and reduced them
to a shorter growth ; though it is said that when the
land is permitted to lie idle under fence they recover
their old luxuriance. We have seen a species of wild
timothy growing four or five feet high on the Tualatin
Plains, in Washington County.

The lives of the early settlers of Oregon, though
not luxurious, were easy and care-free. The genial
climate and the kindly soil rendered constant or excessive labor unnecessary. If they were stock-raisers,
comparative wealth was easily attained, when one hundred cows were worth ten thousand dollars. To mount
his horse, and ride about to look after his cattle, was
a pastime for the stock-raiser ; good riding, good shooting, and knowing how to throw the lasso, popular
accomplishments. Clad in his buckskin suit, and
booted and spurred in true *vaquero* style, it was his
pleasure to scour the prairies day after day on any errand whatever. And well it might be—unless some
of his wild California stock "got after him," when a
sharp race was sure to ensue, which not unfrequently
ended in the herdsman being "treed."

This free-and-easy life, in a country so beautiful, had
many charms which are easily understood. Nor is the
Oregon of to-day so densely populated as to be without
much of the same romantic freedom. Although most
of the open or prairie land in Western Oregon is owned
by donation claimants, locators, and others, comparatively little of it is cultivated. The uncultivated
prairie lands, together with the half-wooded bench

lands of the foot-hills, make a large extent of country still in its primeval condition as to cultivation. One may ride, barring occasional fences, almost at his pleasure from one end of the Wallamet Valley to the other; though for greater convenience he would probably keep to the traveled road. A very pleasant ride it would be, too, if he were fond of equestrian exercise, and by far the best method of obtaining correct notions of the resources of the valley.

It seems at first a remarkable condition of things, that a population of 82,000 people should have appropriated the largest portion of the agricultural lands of Western Oregon—a country 275 miles by forty or fifty in extent. But a large proportion of the open prairie lands were taken up under the Donation Law, which gave 320 acres to a married man, 320 more to his wife, and the same amount to every white male citizen, widow, or head of a family, who would occupy the same according to the requirements of the Act. It is reported of the Oregonians that while the Act was in force, very early marriages were the fashion; and even that the courting which preceded it was sometimes accomplished at the door of a farmer's house, while the would-be husband sat on his "cayuse," and the not unwilling bride of thirteen or fourteen summers stood on the door-step—the object of both being to secure a partner in a mile square of land. Large families who "took up" in this way adjoining "miles" were able to call whole townships their own

So much land, though gladly accepted from Government as compensation for the toils, privations, and dangers of first settling the country, has proved any thing but a blessing to the owners, by preventing close settlement, and the efficient working of a free-school

13

system in the farming districts. Many a farmer has sold his land, where it was somewhat remote from the town, for a merely nominal price, and gone to reside where he could send his children to school. It was impossible, as it was useless, to cultivate a mile square of land, where neither rail-car nor steamboat ever came to take away its produce. And as for stock-raising and wool-growing, it was not necessary to *own* large bodies of land, since there was vacant land enough, of the best kind for that purpose, in the foothills on either side of the valley. If every farmer burdened with a mile square of land had been able to give away half or two-thirds of it to good, intelligent farmers, who would immigrate to take possession of and improve it, the mere fact of their neighborhood to himself, and their assistance in all kinds of enterprises, would so enhance the value of his remaining half or one-third as to make it equal, in value, to the whole, unimproved and isolated.

Doubtless this view of the subject will present itself to the land-holders of Western Oregon, when the lands of the Railroad Companies begin to be sold ; and although they may not wish to give them away, except where they have subscribed for the building of roads, they will be desirous of putting their surplus land into market at very reasonable rates.

We will suppose that we had set out to take a ride through the Wallamet Valley. Starting at the northern end, on the west side, we should take a look at the so-called Tualatin Plains of Washington County. Immediately upon entering them from the heavily timbered Columbia or Wallamet highlands, we are struck with the beautiful natural arrangement of the plains and groves. Small prairies, from one to six miles in diameter, are separated by belts or groups of fir and oak intermingled. Growing in more open spaces than the forest affords, and in a soil of great richness, these trees have attained perfection in size and form. Never have we beheld more truly "Arcadian" groves. It strikes us as a sort of profanation that the farmer at whose house we stop, has allowed one of these grand forest cathedrals to be used as a shelter for his stock, and so to become defiled. Indubitably this is not a utilitarian, nor even a humanitarian view; and the farmer showed care for his cattle, where we should have shown care for the trees. Yet, were not sheds good enough for creatures that are born and die in half a dozen years? and should they be allowed to bring to grief these giants of centuries old?

This county is one of the oldest-settled portions of the State, as the farming improvements show. A large surplus of grain is raised annually, which is wagoned to Portland, or Springville, and there shipped to California, the Sandwich Islands, or some port on the South American coast. The West Side Railroad will soon put an end to the wagoning of grain, and will revive the cultivation of fruit, which has been discontinued on account of the cost and loss of transporting it to a market.

One of the pests of Oregon farming is a large, coarse fern—compound or branching *(Pteris aquilina)*—which is common to the forests, and which encroaches on the improved lands contiguous to them. It is very difficult to eradicate, the roots penetrating to a great depth, and being very tough and strong. Wherever it is found, however, the soil is sure to be good, and more especially adapted to fruit than the exempt prairie. Fern troubles the farmer on the Tualatin Plains, in those fields which border on his timbered land; but thorough plowing and harrowing, or mowing when it is full of sap, will finally kill it.

There is no lack of excellent water in this county. Streams and springs abound; but wells are in general use for domestic purposes, the water being soft, pure, and cold, which is obtained by digging. The Tualatin River is navigated by a small steamer nearly to Hillsboro, the county-seat. Other streams in the foot-hills furnish abundant water-power and mill-sites, which are, in many cases, already occupied; and yet a fresh influx of population would create a demand above the present supply. So nicely is supply and demand adjusted in the farming districts, that there can be no rise and fall of the markets from excess or diminution of current manufactures. This leaves openings for immigrants to begin business, to about the same extent as if the country were entirely new; while the temporary assistance afforded by the older establishments to new settlers greatly lessens the hardships of starting anew.

The price of these "broad acres" in the Tualatin Plains, whose smoothness attracts us, is fully as great as any land in Oregon—being held at from ten to twenty-five for improved, and from three to five dollars for unimproved. In giving the prices of land, allowance for the rise consequent on railroad enterprises will have to be made by the reader; as some parcels, lying along the lines of the roads, or near railroad towns, will increase considerably in value during the current year. The railroad lands will be mostly taken in the foot-hills, where there is a mixture of valley and hill land —small prairie spots, and larger tracts of timber. They will be excellent in quality; of greater variety than the prairie, and better adapted to fruit-growing or the pasturage of stock.

Washington County has Columbia County, to the north, between it and the Columbia River; the Coast Mountains, to the west, between it and the sea; and a high ridge dividing it from Multnomah County and the Wallamet River. South of it, and separated from it by the Chehalem Mountains, lies the famous County of Yamhill. There are probably 350,000 acres in Washington County, of which about one-sixteenth is under cultivation, and five-sixteenths timber.

Hillsboro, the county-seat, is a small and quiet town on a branch of the Tualatin River; not notable, nor particularly handsome in its location. Forest Grove, six miles south-west of Hillsboro, is, on the contrary, beautifully located, near the base of a mountain spur, and is a thriving place, with an academic air. Forest Grove is the seat of the Pacific University—a college under the patronage of the Congregational Church. The present buildings, three in number, are of wood, sufficiently commodious to accommodate the present wants of the country. The Professorships are all filled with men of ability, and the University Library is a valuable one. This college first conferred the degree of A. B., in 1862, upon Mr. Harvey Scott, the present chief editor of the *Oregonian* newspaper, who has kindly furnished us the following notes on the university: "The project of establishing an institution of learning at Forest Grove can scarcely be said to have had its origin as a missionary enterprise, as was notably the case with the educational work at Salem, under direction of the Methodist Episcopal Church, which developed into Wallamet University. Nevertheless, men who came out to Oregon as missionaries, as early as the year 1840, were the men who devoted themselves to the work of building up the institution at Forest Grove. Its founder and most generous patron was Rev. Harvey Clark. He was not a man of wealth, but he was a man of industry, and a man who had thorough ideas of educational work. By donations of land and by vigorous effort among the people, he succeeded in founding an academy, which became quite prosperous. This continued in operation for some years, and attracted much notice as a useful school. In the year 1851, Rev. S. H. Marsh, then a young man from Vermont, came out to Oregon to engage in educational work. He went to Forest Grove, and by his efforts a new era in the institution was commenced. He devoted himself assiduously to his undertaking; the name of "Pacific University" was given to the institution, and it began to make advancement. The academy was continued. College labor devolved almost wholly for the first few years on Professor Marsh, but he was enthusiastic and untiring. Subsequently, Rev. Horace Lyman, who had come out to Oregon about the year 1850, became connected with the university as Professor of Mathematics, which position he still holds. In 1859, Mr. Marsh went to the Eastern States, and succeeded in raising a large endowment for the institution. On his return, its prospects were quite promising, and another Professor was added to the Faculty. Again, in 1868, Mr. Marsh went East, and succeeded in increasing the endowment. At this time, Rev. Geo. H. Collier became connected with the institution as Professor of Natural Science. Three Professors, besides the President, are now engaged in the university, and the endowment is ample for their support. The religious influences are

Congregational, but the institution is not a denominational one. It stands on an independent basis, embodying clear and pronounced educational ideas of its own. It is attaining a growth as one of the distinctive institutions of Oregon, and its prosperity seems assured." In connection with the university is an academy for young ladies, some of the students of which also take the college course.

At Forest Grove, reside some of the earliest settlers of the valley—persons who have seen their children, born in Oregon, grow up to manhood's estate, and have sent them back to "the States" to learn something of an older civilization than that of the mountains and plains of Washington County. There is always a great charm in hearing the annals of a State from the lips of its founders. Many walking cyclopedias of Oregon history belong to the population of the Tualatin Plains, and to their influence is due much of the good order and good morals of the community.

Traveling south from Forest Grove, we soon cross the northern boundary of, and find ourselves in, the beautiful County of Yamhill. Comparisons between counties in this portion of the State, would truly be invidious. Their comparative merits must be very nearly the same ; yet this is, if possible, a more beautiful section than Washington County. Yamhill is, also, one of the first-settled and favorite sections of the valley, with perhaps a little larger population, and a little more cultivated land, than Washington County. It contains eight towns, none of them of much size. Lafayette, the county-seat, is situated on the pretty Yamhill River, about eight miles from its junction with the Wallamet, at the Yamhill rapids or falls, on the north side. A short distance below, and on the opposite side, is Dayton, the grain-depot of the county; and about the same distance above is McMinnville. These are the three principal towns. McMinnville is a handsomely located place, and will be a railroad point. It is the seat of a flourishing academy, as well as the centre of the agricultural interests of the county. The *West Side*, a paper devoted to the interests of the county, is published here.

The Yamhill River is formed by two streams, both rising in the Coast Range, and uniting about ten miles above its mouth. The Salmon River rises in the same pass of the Coast Range through which flow the headwaters of the South Yamhill, but runs in the opposite direction, and empties into the sea. The gorge of these streams furnishes an opening for the sea-breeze to cool the temperature of summer, or moderate that of winter. It is also a roadway for horses and carriages, by which the summer travel reaches the sea-coast. The sea-beach at the mouth of the Salmon River, is a favorite resort for the people of the central portion of the Wallamet Valley. To come here in July, camp out for two or three weeks, fish, ride, hunt, and eat "rock-oysters" and blackberries, is thought to be both a sanitary and a pleasurable manner of taking the summer's recreation.

The "rock-oyster" of Salmon River is so called because it is found embedded in sandstone-rock, and has to be released from its captivity by hard blows with a hammer. *When* it was so encased is not very well known, and the subject is one of no little interest. The quality of the testacean does not seem impaired by confinement; on the contrary, it is said to be remarkably good. The oyster, when extricated from the rock, is pear-shaped, with the impression of a scal-

loped shell on the broad base of the shell which in-
closes it, this being rather soft and tender. At the
small end, or where the stem of a pear would be
placed, is a foot, or feeler, projecting, not only out of
the shell, but also reaching out through an air-hole in
the stone, and probably used to secure food. These
oysters are found at several points along the coast, but
never above the reach of tide-water.

CHAPTER XVII.

THE agricultural capacity of every part of Oregon is so much greater than its present productiveness, that, to state the latter, would only be to disparage the former. It has been estimated that Yamhill County might produce 6,000,000 bushels of wheat, annually; whereas, it actually does produce perhaps one-ninth of that amount. But the time has not yet quite arrived when both the motive and the ability exist for Oregon farmers to do their best.

Yamhill County has produced some of the best stock ever exported from Oregon—the market for it generally being San Francisco. A good deal of stock is annually driven to the Sound in Washington Territory, where it either finds a market, or is exported by water to Vancouver's Island. It used to be that cattle and sheep were raised in the Wallamet Valley for the supply of the mining districts in Eastern Oregon and Idaho, and were shipped up the Columbia River to the Dalles, and thence driven to their destinations. However, since the settlement of the valleys of Eastern Oregon and Idaho, and the fertile Territory of Montana, they have been able, with the help of Utah, to furnish beef and mutton to the miners. Western Oregon still finds a market east of the mountains, but not to so great an extent as formerly.

Yamhill is so peculiar a name, that, to most persons,

it suggests the probability of its being a yam-growing country. The original name, let it here be stated, was Che-am-ill—the Indian term for bald hills—and was applied first to the river at the falls, just above which was the ford, because these hills served as a landmark by which they easily found the ford. The name, corrupted to Yamhill, was bestowed upon one of the counties established under the Provisional Government; and though not particularly euphonious, is distinctive, and in Oregon annals notable.

Crossing the beautiful Che-am-ill hills, we have a charming view of the country on every side, and see again the familiar peaks of Mounts Hood, St. Helen, Adams, and Jefferson. We take leave here of the level plains of Washington and Yamhill counties, and find ourselves among the beautiful, fertile, rolling hills and alluvial valleys of Polk County. This county is about twice the extent of Yamhill, with not far from the same amount of cultivated land, and a few hundreds less population. There are no large towns in Polk County, the people being almost exclusively agricultural. The county-seat, Dallas, is a small place situated on the Rickreal (corruption of *La Creole*) River, nearly opposite to the State capital.

In riding over this lovely section of the Wallamet Valley, the freshly imported Eastern farmer must be struck with the general air of neglect and improvidence. He can not but look with wonder and regret at the shabby farm-houses, the unpruned orchards, and dead-and-alive aspect which pervades the country. Not understanding, perhaps, that which is explainable, and more or less excusable, in this "shiftlessness," he is led to doubt the advantages of the country for farming. But to do that, is to err; the real explanation lies in a knowledge of the early history of the country. In the first place, the farming community of the country was derived originally from the border States, as they were thirty years ago. They had never been *good* farmers in the States of Missouri, Illinois, or Kentucky. Upon immigrating to Oregon they received a large body of land—too large to cultivate properly—with no adequate market for its productions, if they could or would work it. They consequently fell into the habit of raising a little grain indifferently well, of raising stock in the same manner, without caring to improve it materially; of living on what they could buy with the money obtained for what they had to sell—instead of producing butter, cheese, choice fruit, soap, candles, and a hundred things which the careful and thrifty farmer supplies himself with. Of course, this style of farming never improves itself, but constantly grows worse as the years accumulate. The buildings, fences, fields, and farming implements grow constantly more and more dilapidated, while their owners follow suit.

Some of the most beautiful portions of Western Oregon are under this curse of bad stewardship. We have occasion to wonder, when the annual returns are made of so many bushels of grain raised, and so many boxes of fruit shipped, and so many head of fat cattle exported, that they *are* so many. It is because the country is very hard to spoil, that it suffers so little by mismanagement. Not that Polk County, which originated reflections of this sort, is the only, or the chief, offender. There are just as bad farmers to the north of it, and to the south; yet, that there must be some tolerably good ones, the reports of the State Agricultural Society prove beyond cavil.

Polk County has every thing to make it rich and prosperous. All of its prairie and level land, and much of its upland, produces large crops of wheat, barley, and oats. Perhaps four-fifths of the whole county might be turned to grain-raising. It is an excellent fruit-growing region, producing apples, pears, plums, cherries, quinces, and small fruits in perfection. Every garden vegetable produced on this soil is excellent. The grain that does not do well as a crop, is Indian corn; the fruits that fail as a crop are peaches, grapes, apricots, and the like tender varieties. These may be raised in certain localities, but are not sure every year, like the first-mentioned kinds. Eastern Oregon, which is not exposed to the sea-wind and fogs that give to the Wallamet Valley its cool nights and copious moisture, must furnish Western Oregon with Indian corn and peaches.

From the rolling surface of this county, it is evident that good water must be abundant, and mill privileges easily obtained. The mountains furnish plenty of timber for lumbering purposes; the valleys furnish cabinet-woods; and the long, sloping hill-sides are dotted with handsome groves of oak. The mineral resources of the county are as yet undeveloped, but promise to be valuable when they are opened up. Of mills, there are nine which make lumber, and four which manufacture flour, besides one woolen-mill. Schools in the different districts, and academies in three of the towns—Dallas, Bethel, and Monmouth (the latter a college)—evidence the prevailing desire of the people for education.

Next south of Polk, on the west side of the Wallamet River, is Benton County, containing nearly a million acres of land, extending from the river across the

Coast Mountains to the sea. The eastern portion of it, along the Wallamet, is open prairie; while the western is first rolling, then mountainous. All that has been said of the other grain-raising sections applies equally to a considerable portion of Benton County; although this county is more celebrated for fine stock than for any other product. Wool-growing is one of the special interests of Benton, for which its grassy hills particularly adapt it—as also for the dairy business.

In the future development of the country, Benton County should rank high as a manufacturing district; for, besides the woolen factories it is capable of supporting, the lumber-mills it can supply from its mountain forests, and the flouring-mills its grain-fields can keep running, it has extensive beds of coal near the coast in localities where various other manufactures can be carried on, convenient to shipping points.

Perhaps the coast side of the county may sometime be reckoned most valuable for these reasons; and on account of the cod, salmon, and oyster fisheries. A wagon-road from Corvallis, the county-seat, to Yaquina Bay, gives this county an advantage over others that are quite cut off from the sea-coast by the inaccessibility of the mountains. The best dairy-lands in Western Oregon are those creek-bottoms and tide-lands along the coast, where the grass is perpetually green and of excellent quality. Yaquina Creek and Alseya River are two streams rising in St. Mary's Peak, and flowing—the first into Yaquina Bay, and the other into the ocean.

The Alseya really falls into a bay, into which vessels of light tonnage can come in fair weather. The immense cedar forests which border this river make this an excellent point for establishing lumber-mills. The greater portion of this land is still Government land, with the exception of a small Indian Agency: another feature in favor of the coast side of Benton County. The hunter and trapper may find plenty of amusement and occupation about the bays and streams and in the Coast Mountains. Such game as elk, bear, and deer, are plentiful; while water-fowl, beaver, otter, and mink, are more than abundant. Corvallis, the shire-town, has already been noticed in another chapter; besides which there are six or eight smaller towns in this county—ten post-offices in all.

Lane County is the largest county in the Wallamet Valley, with a rare combination of agricultural and manufacturing facilities. Extending, as it does, from the Cascade Mountains on the east, to the Pacific Ocean on the west, and embracing within its limits the three forks of the Wallamet River, besides that branch bearing the *sobriquet* of Long Tom—having thousands of acres of the best grain-land—thousands more of excellent pasture—thousands more of splendid timber, with water-power in abundance—it contains within itself the resources of a small State; being, in fact, more than twice and one-half as large as Rhode Island.

We have already spoken of this county rather particularly in describing the advantages of Eugene City—which must become, to a great extent, a depot for the productions of the upper half of the Wallamet Valley. To the eye, Lane County presents a very attractive diversity of surface: prairies, that from level become undulating; and hills, that from being long swells of scantily wooded uplands, rise gradually into high mountains, with crowns of rugged, evergreen forest.

The value of taxable property in this county is greater in proportion to the amount of land cultivated, than in any other except Multnomah, in which Portland is situated.

The climate of this portion of the valley is rather drier than at the northern end; owing, perhaps, to its greater elevation of four hundred feet. The nature of the soil does not vary much from that of other portions of the valley in similar situations. It is a beautiful sight to behold the luxuriant wheat-fields about the last of June, just before the grain begins to ripen, and when the lovely spotted white lily—*Lilium Washingtonium*—stands head and shoulders higher among it, scenting all the air with its sweetness. The flowers of summer, and the richer landscape tints of autumn, make these valley-pictures always beautiful, sometimes exquisitely so.

Having reached the head of the Wallamet Valley through the counties above named, we find, on returning by the east side, that the principal difference between those on the west and these on the east side of the river, consists in the latter possessing a larger proportion of level prairie-land. There is also rather a better style of farming on this side of the river; on the average, more grain being raised to the acre, and other products in proportion.

Linn County has an area of nearly three thousand square miles; a population of nearly nine thousand; and pays taxes on $3,000,000 assessable property. The estimated productions for the year 1868 were— Wheat, 398,336 bushels; oats, 596,790; corn, 18,084; barley, 11,156; potatoes, 595,790; apples, 107,922; tobacco, 19,108 pounds; wool, 264,296; butter, 526,266; cheese, 8,852; hay, 3,776 tons It was also estimated that this county contained twenty thousand head of cattle, eight thousand horses, twenty-five thousand hogs, and more than fifty thousand head of sheep. The amount of land brought under cultivation within the last three years, must have greatly increased the products of this county; and we regret not being able to give the report for 1871.

It strikes one, on learning the number of cattle, horses, and sheep, that the amount of hay raised is very inadequate to the demand. But the discrepancy is explained by the fact that sheep require no fodder in ordinary winters, when there is no snow. One month of feeding suffices for cattle—should the winter be severe, two months. Farmers sometimes have hay three or four years old before it is necessary to feed it out.

The freedom from care about their stock, proves, as might be expected, occasionally, a snare to the over-confident or negligent stock-raiser; and a winter of unusual severity, with snow, comes to deprive him of the cattle he was too improvident to furnish food for. A month or six weeks of pinching and starving will strew with the carcasses of his cattle and horses those bountiful pastures, which for years had never refused them support. That such a thing should occur is a reproach to the farmer, who has no excuse for not having food enough for a "hard winter" in Western Oregon. The straw which is wasted by burning, if saved, would suffice to feed his stock, in case of need. If not needed, it could be left to rot, and be returned to the fields as manure.

Even in the mildest winters, cattle, especially milch cows, would be much better for foddering; because from the almost constant rains, the grass is watery, and the cattle are drenched and cold. More sheds, and more dry food, would make the cattle and sheep better-looking in the spring ; whereas with the present system they present a rough and miserable appearance by the time the winter is over.

The wealth of Linn County is not confined to its agricultural resources, though its people, with rare good sense, prefer to think so. That part of the Cascade Range in which the Santiam River has its rise is known to produce gold and silver, and also lead. But for reasons easily understood by the reader of our notes on the forests and mountains, "prospecting" is exceedingly difficult in the Cascades ; and probably many years will elapse before the mineral wealth of Western Oregon is even partially understood. It is stored away in the hidden recesses of the mountains, there to remain a promise of employment and riches to future generations, when the population of the Wallamet Valley has become dense enough to drive men into other than the peaceful pursuits of agriculture.

Linn County has a large proportion of prairie-land, with here and there a group of hills, or a single isolated butte, furnishing a pleasant relief to the eye. Besides Albany, the shire-town, it has half a dozen small towns, all pleasantly located and prosperous. It contains fifteen saw-mills, and eight flouring-mills, besides one woolen-mill, one tannery, and several wagon and machine-shops ; and is, in fact, one of the very best divisions of the Wallamet Valley.

Marion County is neither quite so large as Linn, nor has it as great an extent of level prairie ; its surface being more diversified. In fertility it is quite equal ; and in population and property exceeds its southern neighbor. Its mineral and commercial advantages are the same. Its manufacturing establishments are fifteen saw-mills, ten flouring-mills, one pork and beef-packing establishment, one woolen-factory, two carding-machines, one oil-mill, two tanneries, three machine-shops, one foundry, three sash and door factories, and three cabinet-shops.

Probably not more than one-eighth of the land in Marion County was ever broken by the plow. A considerable portion of it belongs to the School-fund, and may still be purchased for two dollars an acre. The prices of farming lands range from three to twenty dollars; and in the vicinity of Salem they command twenty to fifty. The Marion County assessment—indebtedness off—is $3,975,199, an increase of $438,864 over last year. The whole tax in the county this year, including the four-mill tax for building a Court House, is seventeen and a half mills.

Taking Marion as a specimen county, as from its diversity of soil we might, we find, first, that the soil of the river-bottoms is composed of sand, vegetable mold, and various decomposed earths; a new deposit being made, annually, by the winter overflow. These alluvial bottoms are exceedingly fertile, and adapted to corn, tobacco, potatoes, and vegetables of all kinds. Second, the soil of the prairies consists of a mixture of sand loam and alluvial deposit, with a base of clay. It is particularly adapted to the production of all kinds of grain, and tame grasses; and almost equally to roots, vegetables, and fruit. This soil is mellow, and not much affected by drought. Third, the hill land is of a red color, much impregnated with iron, in the form of a black sand, such as is found in the gold placers of Southern Oregon and California. There is also alluvium mixed with this earth, being the wash of the mountains. This soil is excellent for almost any purpose, producing superior wheat, and being better adapted to fruit than the soil of the prairies.

It is a pity that agricultural societies have not thought of giving prizes for model farms. It would be gratifying to know just what the land of Marion County, for instance, would produce, if made to do its best. We find at fairs choice lots of wheat, or oats, which may be the result more of accident than of good farming. We hear many persons say that twenty bushels of wheat to the acre is a fair crop; and others who profess to raise sixty bushels to the acre. Somewhere between the two extremes is the mean product of well-tilled land.

There certainly are some farms which yield fifty bushels to the acre, of wheat weighing sixty-six pounds to the bushel; and oats eighty bushels to the acre, weighing forty-seven pounds to the bushel. If any of these farms can be made to produce this amount of grain year after year, then we shall know what the Wallamet Valley can do toward provisioning the world. But nobody knows what is the greatest capacity of these farms, because almost nobody ever does any thing to improve or to restore the land. Twenty years of grain-raising, without manuring, has been wearing out the oldest land instead of improving it.

We have been assured that nine-tenths of the winter wheat raised in the Wallamet Valley has been sowed in February or March, on ground that had been plowed when saturated with the winter rains, and harrowed when the only effect of the harrow was to make it lumpy. After thus "mudding in" the seed, a crop of eighteen bushels to the acre is the result; while wheat sowed in the fall always produces a full crop. The ground for wheat, it is said, should be plowed late in the spring, before it has become too dry ; and plowed deep. In September, or when the first light rains come to soften the earth, it should be cross-plowed and sowed. During summer the ground is too dry, and during winter too wet, for the plow. Wheat properly put in, on good ground, and having a whole summer of sunshine, without storm, to ripen in, and a harvest without rain, has every chance of turning out forty bushels to the acre ; and this is probably not too much to expect of Marion County throughout. The farm of Hon. Sam. Brown, adjoining the new railroad town of Gervais, is the crack farm of the east side. It contains one thousand acres under fence, and has under cultivation several hundred acres. A fine substantial farm-house, with all the necessary out-buildings in good repair, give the place an air of age and wealth which few Oregon farms possess.

Marion may also be considered a type of the Wallamet Valley in its other natural resources—of timber, water-power, and minerals ; and, like the agricultural resources, they are scarcely yet touched upon by the hand of improvement. All the varieties of lumber-making trees and timber for cabinet and other purposes, which have been named elsewhere, are native to the mountains, the plains, or the river-bottoms of this county.

Of towns or post-offices outside of Salem, the county has twelve. One of the most thrifty of these is Aurora, on the line of the Oregon and California Railroad. Aurora is settled by a colony of Dutch, who own sixteen thousand acres of land, which they cultivate on Fourier principles ; and suffer themselves to be ruled over by an autocrat, named Dr. Lyle, who manages not only their financial, but their spiritual and material affairs, quite to the general satisfaction. This seems to be just such another colony as that one settled at Zoar, in Ohio—a place famous for peace, plenty, and cheerful industry. They have a common interest, a common religion, and a common political creed— republican.

Clackamas and Multnomah counties are not, to any great extent, grain-growing—both being covered with timber, except some prairie spots. Farms are yearly being cleared out of the timbered land, but oftener for fruits and vegetables than for grain. The quality of the land is excellent, and its neighborhood to manufactures and to commerce will always make it valuable. The timber, water-power, mineral deposits, and fisheries of Clackamas County, seem to point to its future commercial prominence. The woolen-mill at Oregon City, and the iron-works at Oswego, are but the indications of its adaptability to manufactures.

The agricultural portion of Multnomah is comprised in eight miles of level timbered land, between the Wallamet and Columbia rivers; Sauvie's Island, with several other small islands, in the Wallamet; and a strip of bottom-land extending along the river—in all amounting to perhaps fifty thousand acres. The remainder is mountainous and heavily timbered, with occasional meadows, or ancient beaver-dams. It is the richest county in the State, owing to having Portland for its county-seat. On the very northern boundary of the county, adjoining Columbia County, are some valuable salt-springs, from which have been manufactured the very finest quality of salt, but not in quantity to supply the demand for best dairy and meat-curing salt.

There is another division of the Wallamet Valley—Columbia County—which belongs about equally to the Columbia Valley. It borders for thirty-five miles on the Columbia River, and for fifteen on the Wallamet. It has the Tualatin Plains for its southern boundary, and the Coast Range for its western. It contains about two hundred thousand acres of heavily timbered uplands and ridges, and about one hundred thousand of rich bottom-lands—most of it subject to overflow, in the summer flood of the Columbia.

Where the land has been cleared and farmed, it has proven very productive; the farmers preferring to raise fruit and vegetables to grain, and more of them being stock-raisers and dairymen than agriculturists.

The resources of Columbia County really lie in her timber, water-power, iron-beds, coal-mines, fisheries, and salt-springs. Her advantages are rather those of a commercial, than a farming, district. Lying just between the great grain-growing region and the great natural highway of commerce—the Columbia River—it can not be long before her natural wharves of solid basalt shall be in use to accommodate the exchange between these two.

The whole northern boundary of this county has a depth of water along it, varying from forty to seventy-two feet, with a channel wide enough in most places for vessels to "round out" with ease. These advantages can not be disregarded in the planning of the best and shortest routes for trade and travel. Whether or not the North Pacific and Oregon Central Railroad centre at Portland for the present, the time can not be far distant when an air-line road, from the Columbia River to some point on the valley roads, will be constructed; thus making, direct, a line from Puget Sound to San Francisco Bay. The future of Columbia County as a commercial district, will then be more assured than any other in Western Oregon, unless Astoria should finally become the great city of the Columbia; and even then, all the inland trade would drift to the Columbia by the air-line road.

The summer climate of Columbia County is several degrees cooler than that of Multnomah, having the breeze direct from the sea, by way of the Columbia River. In winter the south-west storms do not have access to it with full violence along the Columbia, on account of the sheltering hills toward the south. It has opposite to it some of the richest lands, especially dairy-lands, in Washington Territory; and Sauvie's Island is just at its eastern end. At present the population is small, but well-to-do and industrious. It has six lumber-mills, and one grist-mill, with others in course of erection. The steam saw-mill at St. Helen is one of the largest, if not the largest, in the State. Its capacity has been given elsewhere.

There are several small streams emptying into the Columbia in this county, whose valleys are being rapidly settled up by individuals or by colonies. The Claskenine, in the western end of it, has some excellent farms along its course. The farmers in the Columbia Valley have the advantage of lumbering and fishing, in addition to farming, as a means of acquiring wealth—an advantage which begins to be perceived in the increasing prosperity of this most sparsely settled portion of Oregon.

The entire area of the Wallamet Valley is about that of the State of Connecticut—or five thousand square miles—with almost no waste-land in it. It is entirely surrounded by mountains, except on the north end, and is 125 miles long, by an average breadth of 40 miles; without estimating the mountainous and timbered country, which would more than double the number of square miles. The soil of any portion, even as high as the tops of the Coast Mountains, is fertile; its adaptability to farming purposes depending almost entirely on its situation with regard to altitude or moisture.

CHAPTER XVIII.

THE UMPQUA VALLEY.

IT was a clear, sharp, October morning, when we left Eugene to go down into Southern Oregon. As the stage rattled out of town in the direction of the Umpqua, we took a last, lingering look at the fair, level valley we were leaving. The encircling hills of russet-color, dotted with bits of green, in groups of oaks or pines; of Spenser's Butte, with its sharp, dark-tinted cone; and of the blue Cascades, now purpling under the morning sunrise. From the most distant mountains, light-gray mists were rising; in the middle distance, was a purple interval; on the nearer hills, rich, yellow sunlight. The orb of day was not yet high enough to shine on the hither side of the peaks behind which he was mounting. They stood in their own shadow, and let his slant beams bridge the valleys between their royal heights, until they rested on the humbler foot-hills among which we were wending our way, and touched with a golden radiance the yellow leaves of the maples, or silvered the ripples in the Wallamet water.

. Such gorgeousness of color never shone, out of the tropics, as the vine-maple, ash, and white-maple display, along the streams in this part of Oregon. We had thought them bright, glowing, radiant, on the Columbia and Lower Wallamet; but nowhere had we found them so brilliant as at the head of the Wallamet

Valley. And, as we afterward ascertained, this is nearly the southern limit of the beautiful vine-maple. It was almost in vain that we looked for its scarlet-flaming thickets fifty miles farther south; and at a hundred miles it had disappeared from the landscape altogether.

The Umpqua Valley is divided from the Wallamet by a transverse range of mountains—spurs either of the Cascade or Coast Range, or both intermingled—called the Calapooyas. The road leads through the gorge of a creek, where the thick woods, in places, quite exclude the sun—almost the light of day. Bright as the weather was, and dry as the autumn had been, there was a shadow, coolness, and moisture here, among the thick-standing, giant trees, the under-wood, and the ferns and mosses. A very pleasant ride on such a morning, but one which might be exceedingly uncomfortable in the rainy season, though never an uninteresting one.

The Umpqua Valley, which we had first seen in its June freshness, was now sere with the long drought of summer, followed by a rainless autumn. Still, it looked beautiful—one so soon learns to admire the soft coloring of these dry countries—the pale, russet hues of the valleys; the neutral tints in rocks and fences; the quiet, dark-green of the forests; and the clear, pale, unclouded blue of the heavens. The expression of these landscapes is that of soft repose. Nature herself seems resting, and it is no reproach to man that he, too, forgets to work, and only dreams. But the men of this period are not dreamers. Even in the sacredest haunts of Nature, they plot business, and talk railroad! We certainly *thought* railroad, as our eyes wandered over this beautiful, but isolated valley, and our imagination became busy with the future.

This valley, or Douglas County, covers an area of 4,950 square miles. Unlike the Wallamet, it has no great extent of level prairie-land bordering on the river from which it derives its name, but is a rolling country, a perfect jumble of small valleys and ridges; the valleys prairies, and the hills wooded with fir on top, but generally bare, or dotted with oak, on their long, sloping ridges. It is a sort of country where a man may seem to have a little world all to himself; owning mountains, hills, plains, and streams, or at least a stream; and not either overlooked by, or at any great distance from, a neighbor.

Extending from the Cascade Mountains to the Pacific Ocean, east and west, and bounded on the north and south by transverse ranges, it embraces all the country drained by the Umpqua River; and is in size and resources fit to constitute a State by itself. Its more southern latitude, greater elevation, and climate, with a mingling of sea-breezes and mountain air, gives it many advantages, making it salubrious and productive. Its prairies are adapted to wheat and all cereals; its creek-bottoms to Indian corn, melons, and vegetables; its foot-hills to every variety of fruit; and its uplands to grazing.

The same general variety of timber grows here as in the Wallamet Valley; and a few kinds in addition. The evergreen myrtle is a fine cabinet-wood not known to Northern Oregon; the wild plum and wild grape also are native to this county; and the splendid *Rhododendron Maximum*, with its immense flowers, of a deep rose color. A great variety of wild flowers adorn the grassy slopes in summer. Strawberries of several native species are abundant, and delicious. Game abounds in the mountains; fish in the streams. In this month of October we saw on the apple and pear-trees a new set of blossoms—some of the fruit having grown as large as a gooseberry.

Douglas County has under cultivation only twenty-five thousand acres, with a population of six thousand. From this average it will be seen that grazing is more followed than grain-growing. The reasons are obvious for the preponderance of stock-raisers: the difficulty of getting so heavy a product as grain to market, over mountain roads; and the greater profit of wool, which *can* be exported; or of beef-cattle and hogs, which can be driven to the mines, in adjacent counties, or California.

Douglas County has a sea-port of its own—Scottsburg—situated at the head of navigation, on the Umpqua River, about thirty miles from the sea. Umpqua Bay is a small, but safe harbor, into which vessels and steamers of light draught can come with ease. It was once projected to build up a city at the mouth of the river, and a company for that purpose was formed in 1849; but the project was finally abandoned as being premature. Scottsburg is at present the main *entrepot* for the commerce of this valley—from which port goods are wagoned fifty miles to Roseburg. Late surveys of the river between Scottsburg and Roseburg have resulted in an attempt to improve its navigation between these two places, so that boats can come up to Roseburg about six months in the year.

The resources of the Umpqua Valley, besides its agriculture and stock-raising, are gold-mining and lumbering. Really, its mineral wealth is very little known. Coal beds are found on the north fork of the Umpqua.

Limestone, brown sandstone, salt-springs, besides sulphur and soda-springs, are known to exist; but these things are left untouched until a more numerous population calls for their appropriation and use. Salmon-fishing is carried on to some extent near the mouth of the river; and also fishing for oysters along the coast. The coast country is an excellent one for fruit-raising and butter-making; for the former on account of the absence of frost, for the latter on account of excellent and ever-fresh grasses, cold spring-water, and even temperature.

Traveling through the valley by the stage-road from the north, Oakland is the first town of much importance we come to. It is beautifully situated on a branch of the Umpqua River, among a grove of the trees from which it takes its name. Among its public buildings is the Oakland Academy, used both for school purposes and for the holding of religious services by the Methodists; and a Masonic Hall. Ten miles south of Oakland is another academy—at Wilbur—also under the patronage of the Methodist Church.

Roseburg, the county-seat, is a pretty little town of five hundred inhabitants, charmingly located in one of the oak parks bordering the Umpqua River. It has an academy, four churches, a Masonic and Odd Fellows' Hall, and public schools; and all the usual trades and manufactures of an inland town. Too much can not be said of the landscape beauty of this part of the county. It is easy to foresee that when these valleys are made accessible they will be populated rapidly; as well from their attractiveness as from the excellence of soil and climate. A United States Land Office is located at Roseburg. Farming land can be purchased at from three to fifteen dollars an acre.

We have already referred to wool-growing as one of the leading interests of Douglas County. In the year 1869 this county exported 430,000 pounds; and the amount is annually increasing. The same year it contained 11,000 head of cattle and 160,000 sheep. The amount of bacon exported is not known, though it is considerable. The oak glades of Umpqua furnish great quantities of food for hogs, at no expense—the acorns seldom failing to be a good crop.

Water-power, of unlimited extent, can be had—the finest being near Winchester, on the Middle Fork of the Umpqua. The site has been offered as a gift to any Company who would erect manufactories at this place. It is an excellent situation for a woolen-mill, being about in the centre of the county.

The road from Roseburg, toward the south, gives us views of very great beauty and grandeur. Every variety of surface is presented, including prairie; gentle slopes, picturesquely wooded; mountain ridges; wild canyons; and every form of noble or pleasing landscape. The Myrtle Creek Hills remind us of Harper's Ferry. We are awed and delighted with the Umpqua Canyon, fully as wonderful as the more celebrated Echo Canyon. The valley of Cow Creek fascinates us, with its wild and solitary beauty, and the extraordinary richness of the autumn tints with which the mountain sides are resplendent.

The scenery does not fail for one moment to interest the traveler during the long ride which takes him from the Umpqua into the Rogue River Valley, over a range of mountains of that name. Traces of old mining operations begin to appear along this route. The earth is broken and scarred; old, deserted cabins stare blankly at us from the roadside; abandoned rockers

and pans, testify to the hope and the despair, or the
success, of former gold-miners. Passing through a
country where the soil is a reddish clay, clothed with
groves of oak, manzanita, laurel, and pine, we come at
last to the Rogue River, the most beautiful of mount-
ain-born streams. Quite near the river, on the stage-
road, the traveler finds a neat hotel, with garden attached, looking so home-like, in conjunction with its
beautiful surroundings, that the temptation to stop
over for a day, and enjoy the peace and pleasantness
of the place, is almost irresistible—to us, quite so.

CHAPTER XIX.

ROGUE RIVER VALLEY, like the Umpqua, extends from the Cascade Range to the sea; embracing all the country drained by Rogue River and its tributaries. It has the Umpqua Mountains on the north, the Siskiyou Mountains on the south, and is the most southern division of Western Oregon. This valley, like the Umpqua, is an aggregation of smaller valleys, divided by rolling hills, and the whole encircled by elevated mountain ranges. The Rogue River is not navigable any great distance from its mouth, owing to the numerous rapids and falls with which it abounds; but for the same reason furnishes abundant water-power. Ocean steamers can enter and carry freight as far up as Ellensburg. It is a stream of unsurpassed beauty, with water as blue as a clear sky, and banks overhung, in some places, with wild trees, shaggy cliffs, and in others by thickets of grape-vines and blossoming shrubbery.

About half a mile off the road to Jacksonville is a fall one hundred and fifty feet in height, down which the river plunges, between rocky cliffs, into a basin in the gorge below, and then rushes roaring over its rocky bed, for some distance, through a deep and narrow ravine—the whole forming one of the most beautiful of the many beautiful wild scenes in this altogether picturesque country.

15

It is not claimed that there is as great an amount of rich alluvial soil in this section of Oregon as in the valleys north of it. It is rather more elevated, drier, and on the whole more adapted to grazing than to the growth of cereals. Still, there is enough of rich land to supply its own population, however dense ; and for fruit-growing no better soil need be looked for. A sort of compromise between the dryness of California and the moisture of Northern Oregon and Washington—warmer than the latter, from its more southern latitude, yet not too warm by reason of its altitude— the climate of this valley renders it most desirable. Midway between San Francisco Bay and the Columbia River, what with its own fruitfulness, and the productions of the Wallamet and Sacramento valleys on either hand, within a few hours by railway carriage— the markets of the Rogue River Valley can be freshly supplied with both temperate and semi-tropical luxuries.

The grape, peach, apricot, and nectarine, which are cultivated with difficulty in the Wallamet Valley, thrive excellently in this more high and southern location. The creek-bottoms produce Indian corn, tobacco, and vegetables, equally well ; and the more elevated plateaux produce wheat of excellent quality, and large quantity, where they have been cultivated : still, as before stated, this valley is commonly understood to be a stock-raising, fruit, and wool-growing country—perhaps because that kind of farming is at once easy and lucrative—and because so good a market for fruit, beef, mutton, bacon, and dairy products has always existed in the mines of this valley and California.

The placer-mines of Rogue River Valley continued

to yield gold in paying quantities to white men, for about twelve years; since when, the diggings have chiefly been abandoned to Chinamen, who are content with smaller profits. Quartz-leads bearing gold, copper, and silver mines are known to exist in this valley, as well as lead, iron, and coal mines; but the limited capital of the inhabitants, and the greater security of other means of living, have caused them to remain undeveloped.

Like every part of the Pacific Coast, this valley has its mineral springs; and like all the rest of Oregon, its trout-streams, its fine forests, game, and abundance of good, soft water. No local causes for disease seem to exist here; and with care to avoid the miasma always arising from freshly broken ground, we can not conceive of a country more naturally healthful, or in every way pleasant to live in.

The Rogue River Valley is divided into three counties—Jackson, Josephine, and Curry. Jackson County covers an area of 11,556 square miles, and has a population of 4,759; about fifteen thousand acres of cultivated land, and assessable property to the amount of $1,500,000. The price of farming land is from five to ten dollars per acre.

Jacksonville, the county-seat of Jackson County, with a population of one thousand, is located at the head of a valley, forty miles long by about twelve wide, near the foot of the Siskiyou Mountains, in a romantic and beautiful situation. It is a thriving business place, being the point of exchange between the mining and the agricultural population. Ashland, the second town in the county, sixteen miles southeast of Jacksonville, has a fine water-power, and a woolen-mill erected upon it, which manufactures

blankets, flannels, and cassimeres. A flouring-mill,
and two lumber-mills, are also located here; besides
a marble-factory and machine-shop — showing the
manufacturing enterprise of a small community. The
marble used here is taken from a quarry close by, and
is of a good quality. It is sparkling, white, hard, and
translucent; looking like a conglomerate of large
crystals. It is sawed by water-power, the saw only
penetrating about three inches per day.

Josephine County embraces 2,500 square miles of
the more mountainous middle division of the Rogue
River Valley. Only about six thousand acres have
been put under cultivation. Its population is dispro-
portionately large, when the amount of land cultivated
is considered; which only proves that its principal
wealth is presumed to consist in its mines of gold,
silver, and copper. Mining has been carried on with
profit for about ten years; and the enterprise of some
companies in turning the water out of the beds of
some of the streams, has lately opened up rich placers
of gold, and given a new impetus to gold-mining.

Copper-mining has not been so successful, chiefly
on account of the purity of the metal, making it difficult to work. Another obstacle, is, want of transportation for the ore to any port or shipping-point. This
latter obstacle to mining operations is one that time
and capital will remove. The chief mining localities
are on Josephine, Althouse, Sucker, and other tributary creeks flowing into the Illinois River, itself a
tributary of Rogue River.

Owing to the shifting nature of mining populations
everywhere, Josephine County has less assessable
property than other portions of the country. Yet it
is one of the most delightful parts of Oregon, with

grand mountains and quiet, fertile valleys, lying between beautiful slopes ; with oak groves looking like
old orchards, and open woods of the noble sugar-pine;
with abundant wild fruits and flowers, balmy airs, and
odors of sweet-scented violets. "It is," a lady said
to us, "a paradise of beauty, where, if one had one's
friends, life would be as charming as could be desired."

Kirbyville is the shire-town of Josephine County,
situated on the Illinois River, and doing the business
of a flourishing country town. Several other places
of minor importance are located on the different
streams. Educational and religious privileges have
not kept pace with other improvements in this part of
the Rogue River Valley, for the same reason that renders all mining localities inattentive to such matters—
the want of a permanent population. They wait for
an influx of steady-going settlers with families, a great
number of whom could find delightful homes in Josephine County, at Government prices, or under the
homestead law.

Curry County differs from Josephine, in being more
heavily timbered, as the mountains nearest the coast
are always found to be. In among the mountains are
some small prairies, and others are found extending
along the sea-shore. The soil everywhere is highly
productive ; but owing to the great preponderance of
lumbering and mineral interests, this county will not
become notable for agriculture, though it might be
esteemed an excellent fruit or dairy country. Its population is small, on account of its inaccessibility. The
present population follow gold-mining, chiefly on the
ocean-beach, where is an inexhaustible mine, which
the winter winds and tides throw up each year for the

work of the following summer. The gold which is everywhere found on the coast of Oregon, but more particularly this southern portion, conclusively proves that deposits of the precious metal exist in the Cascade or Coast mountains, or both. That which is found at the mouth of the Umpqua and Rogue rivers might have been washed from the Cascade Range, as those rivers rise there. But farther north, on the coast, where the streams all rise in the Coast Range, gold is also found, though it has not been mined, as in these localities it has. In fact, the "color" may be "raised" in almost any stream in Oregon, and we have seen it taken out of the gravel in a well which was being dug in Portland.

Curry County is well supplied with game and fish. Its splendid cedar forests are worth more than a gold-mine to whoever will convert them into lumber. Cedar-trees that have not a limb on them for a hundred feet, and from three to eight feet in diameter, are not uncommon. Port Orford, the only port of the Rogue River Valley, is in this county, and also Cape Blanco, the westernmost point in Oregon. There is good harborage at Port Orford, and water enough for such vessels as are used in the lumber trade. In fair weather, the ocean steamers sometimes call here. A road is built across the mountains from the port into the Umpqua Valley; so that, with some improvements, Curry County might be brought into note for its natural productions, instead of being considered too far out of the world to be habitable. Ellenburg is the county-seat.

Curry County shares, in common with all the coast country, a climate superior in some respects to the valleys. The changes in temperature are less than in the interior; being cooler in summer, and warmer in winter. The sea-fogs keep the vegetation forever green; and miasmatic diseases are unknown. These are certainly advantages not to be contemned. The settlers in the valleys would like to live on the coast, if it were not for the mountains between it and their fertile prairies. Yet, it is just by these mountains that the climate of each division is made what it is— partially confining the sea-fogs and winds to the coast, by which one is made cool and moist, while the other is comparatively warm and dry.

CHAPTER XX.

Lying north of Curry, is another coast county—small, but well known, and of commercial importance—and that is Coos. It has a population of about seventeen hundred; and Empire City, the county-seat, has nearly five hundred. This county is famous for its coal and lumber. Coos Bay coal is well known in the San Francisco market, the mines having been worked for several years. Several lumber-mills do a large business in cutting lumber for foreign markets; and the business of preserving fruit, by drying, is carried on to some extent. The natural resources of the county are not yet begun to be developed; but they are numerous enough to make it famous for its wealth, whenever they shall be open to trade.

The coast country of Oregon is peculiar in its physical features. The Coast Range of mountains leaves but a narrow strip of country between itself and the sea; and were this narrow belt all of the arable land on the sea-side, it would be but little. But the Coast Range sends down a great number of small rivers, all of which have narrow valleys lying between high, timbered ridges. These valleys are extremely fertile, the soil being composed of the wash of the mountains, mixed with sand, and furnished with abundant moisture. Most generally, the borders of such streams are covered with a dense growth of alder, vine-maple, wild-cherry, and other thrifty, small trees; but in some cases, the bottom-lands widen out so as to leave fine prairie spots between the streams and the spurs of the hills. All these valleys will grow grain, hardy fruits, and the finest of vegetables, in abundance. But owing to the great amount of moisture from the sea, which keeps ever verdant the nutritious native grasses, it is especially a dairy country. The coolness and evenness of the temperature along the coast is another advantage to dairying, together with the great amount of root-crops which the ground produces, of the kinds best for milch cows. But the diversity of surface allows the farmer to choose what branch of farming he will follow: whether it will please him to raise grains, hay, fruit, vegetables, or make butter and cheese.

Very many of the coast streams empty into bays of their own. At the mouth of the Coquille and Rogue rivers are harbors, which have been used to some extent by small vessels—while Coos Bay is the leading sea-port for Southern Oregon. Mean low-water on the bar is eleven feet; high-water, sixteen feet seven inches. Umpqua Bay is more of an open roadstead than Coos; but furnishes a very good harbor, with thirteen feet on the bar, at mean low-water; and nineteen feet, high-water.

The Alseya River forms a small bay at its mouth, which is not much used. Yaquina Bay, however, is quite an important port, where vessels from San Francisco come to load with lumber and oysters. It has a straight entrance, half a mile wide, with fourteen feet of water on the bar at low-tide. The name, Yaquina, signifying *everywhere*, describes the shape of the bay, however, when once inside. It meanders about for nine miles, having no less than three settlements along its shores. Newport, the principal one, is located on the north side of the bay, on the site of an old Indian town—the site being marked by the holes in the ground where stood the ancient wigwams, and by piles of oyster-shells, showing how the tenants of these primitive dwellings lived. At the head of the bay is Elk City, the terminus of the Corvallis wagon-road, which passes through a low gap in the mountains formed by the valleys of two streams, one entering Yaquina Bay and the other the Wallamet River. On the Newport Hills, on the north side of the bay, is a third-class light; and on Cape Foulweather, still farther north, a light of the first class. On the south beach is a strip of sandy plains, covered with a scattering growth of pines, which are singularly dwarfed, bearing cones when not more than two feet high.

The Siletz River is a large and rapid stream, with a valley of considerable extent, in which is an Indian Reservation. It is principally timbered land, with a soil of black muck. The tribes gathered on the reservation are remnants of nearly all the tribes of Western Oregon, from the Columbia to the southern boundary of the State. They are tolerably well taught in agriculture, and seem desirous of attaining to a higher civilization. Tillamook Bay, like Yaquina Bay, is the outlet of a river of the same name. It is a good harbor, with sufficient depth of water on the bar for the passage of light-draught vessels. Tillamook County, in which this bay is situated, extends from Benton County on the south, to Clatsop County on the north; and has five small rivers, flowing from the mountains to the sea. Its population is only about four hundred; and its business is confined principally to lumbering and fishing. It has two saw-mills and two grist-mills.

A large area of land belongs to this county, probably
1,280,000 acres; of which the greater part is covered
with timber. The small valleys we have mentioned,
and slopes of many of the timbered ridges, furnish a
fair proportion of arable land.

The Nehalem River, which forms the boundary between Tillamook and Clatsop counties, although possessing no harbor at its entrance, has one of the most
important valleys on the coast. This river rises in the
Coast Mountains, far to the east, and flows through
them, by long meanderings, to the sea, having its
mouth only a few miles north of Tillamook Bay. For
twenty miles back from the ocean the country along
the Nehalem is broken; but at this distance the valley
opens out, from half a mile to a mile and a half in
width, and continues of this breadth for forty or fifty
miles. The soil is a sandy loam, very warm and fertile. On each side of this valley, or bottom-land, the
country rises with a gentle slope, far back, and is covered with a fine growth of the best timber; the forest
being little obstructed by undergrowth. The soil of
the timbered land is also a rich black loam, of great
depth, which will make fine farms, when cleared of the
timber. The Nehalem country is attracting much attention; and although still unsurveyed, is rapidly being taken up by settlers.

Clatsop County we have already described in an
early chapter. To sum up the coast country of Oregon: it is a narrow strip of country along the sea, often
intersected by small rivers, some of them with bays
of a large size, suitable for harbors; and all of them
with some excellent bottom-lands back between the
ridges. The bottom-lands are generally covered with
a growth of alder. cherry, vine-maple, and kindred

small trees; while the slopes of the mountains, and even the highest ridges, are heavily timbered. The soil is excellent, though generally too cold, when taken together with the moisture of the climate, for producing and ripening the cereals, and tender fruits; but very productive in all kinds of grasses, roots, hardy vegetables, and fruits—such as apples, pears, plums, and cherries.

It is an excellent country for lumbering, wherever there is communication with a market. The streams abound with speckled trout; the bays and inlets with salmon, and different kinds of fish; as well as oysters and clams. Game of every variety is plentiful in the mountains and valleys. Coal, iron, copper, and gold, one or all, are found in every county. Owing to extensive fires, which swept over the Coast Range in 1847, from Tillamook to Umpqua, a large portion of the timbered hills are-comparatively easy to clear. Of the numerous rivers of clear, cold water, several are navigable for small boats, such as run on Yaquina, Tillamook and Coos bays. In short, the coast counties are capable of supporting a large population of mining, lumbering, stock-raising, dairying, fishing, and farming peoples.

The climate of the coast is extremely healthful, from the evenness of its temperature, freedom from miasma, and invigorating sea-air. The height of the Coast Range varies from two thousand to five thousand feet; and the wall thus interposed intercepts most of the fogs of summer, causing them to be precipitated upon these hills and valleys of the west side—the result being coolness and moisture: yet, when winter comes, sometimes with rigor, to the interior, the same causes operate to keep the coast of a milder degree of cold.

To be convinced of the prodigal wealth of the soil and mildness of the climate, we have only to visit any of the small valleys about the mouth of the Columbia River, and find ourselves lost as in a forest, in thickets of ferns and wild berries; while the trees that tower above us two hundred feet have trunks eight feet in thickness. Of the scenery of a country which combines sea, mountains, forests, valleys, and rapid rivers, we need not speak. The imagination has ample scope here; but its variety could never surpass the infinite variety of the scenes furnished by Nature's never-ending new combinations.

CHAPTER XXI.

FROM THE COLUMBIA TO THE SOUND.

SUPPOSING the tourist to have arrived in Oregon by the usual routes of sea or overland travel, he is sure to be carried to Portland; from which point radiation to all other parts of the north-west commences. We take a steamer at that place, and retrace our course to the Cowlitz River; taking six hours for the voyage, which ends at Monticello, in time for a one-o'clock dinner. Here we find one or more stages waiting to convey mails and passengers to Olympia; and if competition be strong, for very cheap fares. Our stage, on this occasion, is a long, light, open wagon, well loaded down with mail matter before we take our seats. The first six miles are along the river-bank, in sand and dust, with very little open country in sight; this portion of the Cowlitz Valley being of no great extent. Then commences the crossing of the Cowlitz Mountains.

What strikes us most in this drive, are the magnificence of the timber on the mountain, and the roughness of the country for a highway. In this July weather it is well enough, jolting through the forest, over roots of giant trees, and into hollows between them; but, in the rainy season, it is a different undertaking. However, the North Pacific Railroad is to cure this evil, in another year, perhaps. *We* are glad that for once we had to come this way. Such a forest as this is something to remember having seen; and fills completely our conception of solemn and stupendous grandeur. Fir and cedar are the principal trees. They stand thickly upon the ground ; are as straight as Ionian columns ; so high that it is an effort to look to the tops of them ; and so large that their diameter corresponds admirably to their height. If there is any thing in Nature for which we have a love resembling love to human creatures, it is for a fine tree. The god Pan, and the old Druidical religion, are intelligible to us, as expressions of the soul struggling "through Nature up to Nature's God ;" and as a religion free from arrogance, and the temptation to build upon itself worldly ambitions, recommends itself even in the nineteenth century. A lover of the woods must enjoy this drive, as we did, both in an esthetical and religious point of view.

It is quite natural in such profound solitudes to look for some of its most distinguished inhabitants ; but our desire to meet a cougar, or a brown bear, is not gratified. Only the gray hare, and the grouse and quail, cross our road. These seem not the least to mind us ; evidently unacquainted with the sanguinary disposition of man, and so audaciously familiar as to provoke an uprising of the lordly thirst for killing. The weather is fine, the mountain air and balsamic odors tonic and delightful. Altogether we are in the best of spirits for two-thirds the distance to the night station. Then growing well acquainted with the scenery about us, we begin to demand fresh excitement, and are rather glad that there is a prospect of breaking down, which requires us to do some walking and some wagon-mending ; so that we arrive at the crossing of Aliquot Creek about dark, and take lodgings at Pumphrey's, on the farther side.

Pumphrey's Landing is at the head of navigation on the Cowlitz.   Until the middle of July a small steamer ran up to this point, but is now discontinued until a return of high-water.   It is from here that canoe passage is taken *down stream* to Monticello—an exciting and pleasant excursion—the river being very rapid, and the Indians very expert.

We are on the road again by day-break, crossing Pumphrey's Mountain before breakfast.   The road, in all respects, resembles that of the day previous.   The morning is quite cool, although it is July weather, and the blazing, open fire which welcomes us at McDonald's, gives the most cheering impression.   Here we obtain a substantial breakfast, and have time to admire the comfortable, home-like appearance of this isolated station.

Our road now lies across McDonald's Prairie, from which we catch the first real view of Mount Rainier, the grandest snow-peak of the Cascade Range ; which fact it pains us to admit, because we had taken Mount Hood to be the highest, and even maintained its pretensions over Mount Shasta, its California rival.   But our eyes convince us that Rainier is chief among the snow-peaks, and altogether lovely.   Measurement makes it just four feet higher than Shasta—so the North has the champion mountain, after all.   The lights and shades upon it, as we catch frequent glimpses during the day, are beautiful beyond criticism.

There is very little good farming land along the line of the road.   Where there is not a thick growth of forest, the intermediate prairies are gravelly, making excellent driving, but poor farming.   Occasionally, where there is a small piece of valley land, it is of the richest description.   The grass that is being cut in some of these little valleys is the heaviest we ever saw.

At the first crossing of the Chehalis River is the pretty village of Claquato, which makes us wonder how it got there, so isolated it seems from the outside world. Its buildings, gardens, and orchards have a truly comfortable, even charming, appearance; and the sign, "Claquato Academy," upon the front of a good-sized frame-building, inspires us with respect for this isolated community. Altogether, it produces a most favorable impression — suggesting numerous quotations from the poets, who, we recollect with a sigh, are not, after all, very reliable real-estate agents.

The Chehalis, near here, receives the waters of the Newaukum, a small river heading in the foot-hills of the Cascades. The valley of the Newaukum, together with that of the Chehalis, above the junction, afford from fifty to seventy-five thousand acres of the best of farming land; Lewis County, which contains them, being one of the best agricultural counties in the Territory. In the Chehalis Valley is a cedar-tree, we are told, measuring twenty-one feet in diameter six feet from the ground, and is estimated to be 250 feet to the first limbs.

We get our last coupon of rough road just beyond Claquato, a few miles of which brings us to the second crossing of the Chehalis, at its junction with the Skookum Chuck ("strong water"), another pretty spot, where we dine. Not more than three miles from here, is a fallen tree three feet in diameter at the butt, and 290 in length. Another tree, in an adjoining county, measures eleven feet in diameter, and 310 in length, and we hear of two more being fourteen feet in thickness; which is pretty well for firs and cedars. From Skookum Chuck to Olympia, is a fifteen-mile drive over gravelly prairies, separated by wooded sections.

16

The Grand Mound Prairie is interesting from the number and regularity of the mounds, which are two or two and a half feet high, and as close together as potato-hills in a field. Various theories of their origin have been assigned, but the satisfactory one has not yet been suggested.

The entrance to Olympia is through a belt of magnificent trees, four or five miles wide. Just at the head of the sound, where the Des Chutes River falls into it, is a little, lumbering village called Tumwater, with a saw-mill, flouring-mill, and tannery. The falls of the Des Chutes are very pretty; but their beauty will ultimately be hidden by all manner of mills, which will be made to avail themselves of this fine water-power. We observe, concerning names, that the river retains its French name for falls, and the town its Indian name for the same thing. Passing through Tumwater, which is but a suburb of Olympia, we soon find ourselves in the streets of this classically denominated capital.

Olympia depends upon its location for its claim to beauty. Like all towns hewn out of the forest, it has a certain roughness of aspect, caused by stumps, fallen timber, and burnt, unfallen trees. But it has decidedly an air of home-comfort and cheerfulness, with snug residences, good sidewalks, and, to us, the singular charm of *long bridges*, and spacious wharves. To be suspended over water on a bridge, a long one, was always to us more fascinating than boating. To be at rest over the ever-restless water, and gaze upon its cheerfulness, and dream! In Olympia, we can do this, when the tide is in. When it is out, we can interest ourselves in watching the millions of squirming things the receding flood leaves in the oozy mud. Standing on the long bridge, too, we can gaze upon the distant Olympian Range—the most aerial mountain view in America.

The following is the history of Olympia in brief, as furnished by one of its citizens: "The land claim on which is situated the town of Olympia was settled on by Mr. Edward Sylvester, in the year 1845. At that time the whole Puget Sound country was a perfect wilderness, excepting the settlements of the Hudson's Bay Company at Nisqually, then in charge of Dr. W. F. Tolmie, and a few pioneer settlers at Tumwater and the prairies south of Olympia, who came in the year before. Mr. Sylvester resided here three years alone, and in 1849 went to the gold-mines in California. Returning early in 1850, he found several new-comers, among whom were three or four families, and shortly after his return he had the town surveyed and laid out. One or two stores were soon started, which supplied several lumbering camps, and the brig *G. W. Kendall*, Captain A. B. Gove, was placed on the route between Olympia and San Francisco, and a profitable business started in furnishing the California market with spars and piles. A large village of Indians was situated along the bank of the bay, on the west side of the town. The road to Tumwater was not open until 1852, and the first bridge was finished the succeeding winter. All the other roads and bridges were later undertakings. The Custom-house District was organized at Olympia November 10, 1851—S. P. Moses, Collector. A weekly mail (horseback and canoe) service from the Columbia River, was first established in 1852—Messrs. Yantis and Rabbeson, contractors. The Down-Sound mails were first carried in 1854 by the steamer *Major Tompkins*, Captain J. S. Hunt; the same steamer was shortly after wrecked while going into Victoria harbor, and she was succeeded for the two years following by the steamer *Traveler*, Captain J. G. Parker. The first newspaper — *The Columbian* — printed north of the Columbia River, was issued on the 11th of September, 1852, at Olympia, by Messrs. Wiley & McElroy. The Methodist denomination had a resident preacher at that time, but the French Catholics built the first church, in 1852. The first school-house was built in 1853, the same being constructed at the expense of, and through the enterprise of, the ladies. The first wharf was built in 1854, by Mr. Edward Giddings. The present site remains the landing of the ocean and Sound steamers. In the fall of 1853, General I. I. Stevens — then recently appointed Governor of Washington Territory and Superintendent of Indian Affairs — arrived overland with his party of surveyors and engineers, then in the interest of the Northern Pacific Railroad Company. In 1853, Olympia was made the capital of Washington Territory."

Olympia, besides being the capital of the Territory, and county-seat of Thurston County, has a most favorable location with respect to the rest of the Territory, being at the head of the Sound, at a point nearest the Columbia River and Gray's Harbor, and about equidistant from all the principal valleys in Western Washington. It has also great advantages in the way of water-power that is contiguous to tide-water. The falls of the Des Chutes furnish, alone, a 1,600-horse power, at the lowest stage of water, and may be made to furnish much more at a slight expense in conducting water from the Nisqually River. Another stream, Percival's Creek, is capable of being made a water-power almost equal to the Des Chutes, by cutting half

a mile of ditching; the same drainage reclaiming
twenty thousand acres of the best grass lands. A
comparatively small expense would build a dock at
Olympia, covering 1,800 acres, in which vessels could
be kept afloat when the tide is out; and such a dock
will no doubt be built before long, whether or not this
city becomes—what, of course, it aspires to be—the
terminus of the Northern Pacific Railroad. The population of Olympia is about thirteen hundred.

Western Washington, unlike Western Oregon, has
no chief river, with its numerous tributaries, draining
a great valley; but it has, nevertheless, its central body
of water, into which flow frequent small rivers, draining the Puget Sound Basin, which is bounded, like the
Wallamet Valley, by the Cascade and Coast ranges, on
the east and west, and by their intermingling spurs on
the south. These rivers, unlike those of Oregon, are
all affected by the ebb and flow of the tides; and have
their lowest bottom-lands overflowed. The Sound
itself is not one simple great inlet of the sea; but is
an indescribably tortuous body of water, which is not
even a sound; being too deep for soundings, in some
of its narrowest parts. So eccentric are its meanderings that the whole county of Kitsap is inclosed so
nearly in the embraces of its several long arms, as
very narrowly to escape being an island.

That particular arm of the Sound upon which Olympia is situated is six miles in length by from one to
one and a half miles in width; narrowing to a quarter
of a mile when opposite the town. At low-tide the
water recedes entirely at this point, leaving a mud flat
all the way from here to Tumwater, a mile and a half
south. The mean rise and fall of the tide is a little
over nine feet; the greatest difference between the

highest and lowest tides, twenty-four feet. Vessels come in on the tide, and lie in the mud to discharge; going out again on the high-tide. The construction of a dock would give them water to lie in of any sufficient depth.

The land adjacent to this inlet is considerably elevated along the shore, and rises yet higher at a little distance back, being level, however, in some places. The same general shape of country surrounds the whole Sound, the land having a general rise back from it for some distance. This, of course, must be the case, where a basin exists of the character of this one. That portion of it which lies adjacent to the Sound possesses a porous, gravelly soil; nevertheless, heavily timbered with trees of immense size. This belt of timber is several miles in width. The roads through it, and across the small prairies which lie on its outskirts, are all that could be desired in the way of natural McAdam, and furnish delightful driving. One thing we observed regarding these beautiful prairie spots, was, that along their edges, where they receive the yearly accession to their soil of the leaf mold of the forest, the orchards and gardens looked very thrifty; and also that wherever there was a piece of bottom-land, on any small stream, the hay-crop was the heaviest we had ever seen.

About ten miles back from the Sound, on the east, the country commences to improve; and from there to the foot-hills of the Cascades furnishes a good grazing region, with many fine locations for farms. The foot-hills themselves furnish extensive clay-loam districts, suitable for grain-raising; and will, when cleared, become very valuable farming lands. Around the base of the Coast or Olympian Range, on the west, there is also another large body of clay-loam land ; and to the south, between the Chehalis and the Columbia—or, more properly, between the Columbia and the higher ground which separates the Columbia Valley from the basin of the Sound—there is a still larger district which may be converted to grain-raising. But the vicinity of the Sound, within a distance of from ten to twenty miles, affords little land that is good for grain, except that of the river-bottoms, and of that only certain portions.

For, as before noticed, these streams coming into the Sound are affected by the tides, the lowest land being overflowed daily. That portion of each valley which is free from submersion furnishes the most fertile soil imaginable for the production of every kind of grain, fruit, and vegetable—if we except melons, grapes, and peaches, which, owing to the cool nights, mature less perfectly than in Eastern Washington. The valleys of these small rivers, like those of Western Oregon, already described, are covered at first with a rank growth of moisture-loving trees, such as the ash, alder, willow, and poplar. But they are easily cleared ; and the soil is of that warm, rich nature, that it produces a rapid growth of every thing intrusted to its bosom. Owing to the fact that these valleys are narrow, and head in mountains at no great distance, they are occasionally subject to floods. As floods never occur, however, except in the rainy, or winter season, a proper precaution in building, and harvesting his crops, should insure the farmer against loss from them when they do occur.

The rivers which empty into the Sound on the east side are the longest, with the greatest amount of alluvial lands. They are the Nisqually, Puyallup, White,

Sikamish, Cedar, Snoqualmie, Snohomish, Stoluqua-
mish, Skagitt, and Nooksahk. Several of these have
two or more branches of about equal importance, and
all of them are navigable for certain distances; the
Skagitt for a distance of fifty miles. This last-named
river rises far to the north-east, in British America,
and flows through mountain gorges for long distances,
like the Columbia and Frazer rivers. Everywhere in
the neighborhood of these rivers, and the Sound, is
timber of excellent quality for lumbering, and in great
quantities. The streams flowing into the Sound from
the west are in all respects similar to those on the east
side, except that they are shorter, and have less bottom-land. They rise in the Olympian Range, and have
but a short distance to flow to reach their outlet.

With regard to the great business of the Sound—lumbering—so much has been written, that more seems
superfluous. In a chapter on Forests, we shall give a
full account of the timbered lands both of Oregon and
Washington, together with the amount of lumber annually produced. It is unnecessary to say more in this
place, than that the shore-line of the Sound is over sixteen hundred miles in length; and that its shores everywhere are heavily timbered, except where fires have,
in some places, ruined the timber. Leaving out all
burnt and unsuitable timber, the amount is still enormous which is excellent for lumbering, and easily
reached. Milling companies buy logs at $4.50 per M.
—the loggers having but short distances to go, and
every facility for hauling at a trifling expense; nor will
they work a piece of timbered land producing less than
thirty thousand feet per acre—the more common yield
being twice or thrice that amount. Water-power is
commonly used only in the small mills, all the large

exporting establishments using steam. Of these large establishments, there are sixteen in Western Washington, fourteen of them being on the Sound.

Although fishing, as a business, has not yet received that attention on the Sound which it has on the Columbia River, it should, in the near future, become a large and profitable trade. Nor would the curing of fish be confined to salmon alone, as on the Columbia it now is. Cod are taken in the Sound, and all along the coast to the north. Vessels are already running from the Sound to the fishing-grounds in the Russian seas, and off the coast of Alaska; and others are yearly being built for this trade. They are brought to the more favorable climate of the Sound to be cured; and the finest cod ever put up on the Pacific Coast have been cured in Washington Territory.

A fish for which the Sound is somewhat celebrated, is the *Eulachon*—a small, but good-flavored fish, of so oily a nature that the Indians dry it to burn for torches: hence it has also been called the candle-fish. The experiment of expressing oil from the dog-fish, for commerce, has lately been tried, we hear, with favorable results. The oil sells readily for fifty cents a gallon to the millmen, as machine-oil. Halibut is common in the Strait of Juan de Fuca, and all along up the Gulf of Georgia, and beyond, through the whole extent of those continuous sounds by which the Northwest Coast is slashed in every direction. Other kinds of fish, good for the table at stated seasons, abound in the waters of the Sound and its tributaries—such as smelt, sardine, oysters, and clams. The speckled trout is taken in every mountain stream, above the reach of tide-water.

We notice in Olympia how abundant are berries,

both wild and cultivated—the blackberry season, in
particular, being at its height. All manner of small
fruits seem abundant, and flowers and vegetables "a
drug in the market." New as the country is, people
seem to live well, dress well, talk well ; and are planning a new empire—and all without any great or exhausting effort.

CHAPTER XXII.

WE start down the Sound on an elegant steamer, called the *Olympia*, very early in the morning, in order to avail ourselves of the tide. It is too early to allow us to study the views which daylight affords ; but we feel assured that blue water, wooded headlands, and fair skies make up the panorama, and that the picture will be continued indefinitely throughout the day. Steilacoom is the first place of any importance we come to, and is really in a most beautiful location ; being situated at the south end of the "Narrows," on a high, gravelly prairie, diversified with groves of fine timber, and gemmed here and there with small, clear lakes, bordered by a scattering growth of round-topped oak - trees. The scenery is unusually fine about Steilacoom, four snow - peaks being in view — Rainier, St. Helen, Adams, and Hood. It is said the finest view of Rainier to be obtained anywhere on the Sound, is to be had at this point. The Olympian Range across the Sound, is another fine feature of the landscape ; while the Sound itself, together with the forests and valleys in sight from here, furnish a middle ground of great beauty.

The harbor at this place is a fine one, with plenty of water, and good anchorage. Steilacoom Creek furnishes a water - power which runs a flouring - mill and woolen mill, with plenty to spare for other manufacturing purposes; being the outlet of a lake situated two hundred feet higher, and only four miles distant. Springs also abound in the same neighborhood, some of them large enough to run the machinery of a mill. Little more than a mile to the east is old Fort Steilacoom, now abandoned. In the building once used as officers' quarters, the insane of the Territory are now confined, having been recently removed from Monticello. If a healthful location and pleasant surroundings can have any effect to "medicine a mind diseased," the location of the Asylum for the Insane is admirably chosen in this instance. The Penitentiary is also located at Steilacoom, but not on the main-land, a small island being devoted to this institution.

Steilacoom has only three hundred inhabitants, and is possessed of three churches and two school-houses. A boarding-school for girls is kept by the Sisters of Charity. The Masons, also, have a hall; the other public buildings being a Court-house and Jail—the latter built of brick, and very substantial. From Steilacoom east, *via* the Nachess Pass, to Wallula, on the Columbia River, is 225 miles. The altitude of this pass is 3,467 feet; but was opened and used as an emigrant road in 1853, and had $20,000 expended on it by the Government, in 1854. It is now, however, so blocked up by fallen timber as to be impassable for wagons. A recent discovery of valuable iron-ore on the Puyallup River, about fifteen miles from Steilacoom, has given additional importance to this place as a manufacturing point. Opposite to Steilacoom, on a small inlet, is an establishment for manufacturing oil from the dog-fish, before spoken of. This establishment is owned by Col. Pardee, an enterprising gentleman from New Haven, Ct.

Leaving Steilacoom, we steam up the "Narrows"—a strait four miles long by one in width, through which the water runs with great force at the ebb and flow of the tide—and pass by Point Defiance, a high bluff on which defensive fortifications may, at some future time, be erected by the Government. A few miles below Point Defiance is Commencement Bay, on which the town of Tacoma is situated—the first of the great lumbering establishments we come to after passing into that wider portion of the Sound known on the maps as Admiralty Inlet; Puget Sound being, in reality, only that portion of this great body of water south of the Narrows.

On the clearing away of the mists of early morning we find the air on the Sound very bright and bracing. A slight breeze just ripples the blue waters of this Mediterranean sea; the summer sky is delicately mottled with flecks of foam-white clouds; seals sport below; birds flit from shore to shore above; a golden silence, only broken by the paddle-wheels of our steamer, wraps all together in a dreamy unreality very charming to the tourist. Occasionally a white sail, gleaming in mid-distance, adds an interest to the scene; while it, at the same time, suggests what these waters will in time resemble, when palaces shall be reflected in their margins, and the winged messengers of commerce shall glide continually from point to point of these now fir-clad slopes, laden with the precious cargoes of the Orient, making this northern sea a second Bosphorus for beauty and magnificence.

Seventy-two miles from Olympia, by steamer, we come to Seattle, the most important commercial town on the Sound. It is situated upon an inlet six and a half miles long by two wide, with a general direction of east by south, known as Dwamish, or Elliot Bay. This inlet has a depth of water through the middle, ranging from forty to eighty-eight fathoms, with from ten to twenty fathoms on the anchoring grounds. It receives the waters of the Dwamish River, a stream which has only a length of ten miles, and is the outlet merely of the Black, Green, White, and Cedar rivers; all of them having rich agricultural valleys, making, in conjunction with the commerce of Seattle, King County the richest county of Western Washington.

Three miles immediately east of Seattle is Lake Washington (or Dwamish), connected with the bay by the Dwamish River. This lake lies but eighteen and a half feet above tide-water, making it a matter of trifling expense to open continuous navigation for small steamers into it. On the borders of Lake Washington, about nine miles from Seattle, is a coal-mine of excellent quality, and inexhaustible quantity. A company is working it, who have barges and steam-tugs on the lake, for its transportation in cars to the tramway which conducts to tide-water. A canal is talked of as an outlet to the lake. Should the United States Government conclude to erect a naval station on Lake Washington, as it may, the question of an outlet suitable for large vessels will no longer be in doubt. Probably no better locality for naval purposes could be selected than this; combining, as it does, fresh water of sufficient depth, exhaustless supplies of ship-building timber within easy distance, and extensive deposits of coal and iron—the latter alone being distant some thirty or forty miles, on the line of the projected railroad through the Snoqualmie Pass.

This famous pass is 2,600 feet above the sea, and only sixty-one miles, in an easterly direction, from Seattle. It is on this ground that this town has so loudly asserted its claim to become the terminus of the Northern Pacific Railroad—a grade of eighty feet to the mile being all that is required to construct the road through the Cascades from this point east. A wagon-road is now open, *via* this pass, to the plains on the eastern side of the Cascade Mountains, and the Yakima Valley. This is the route by which cattle and sheep are driven from the great pastures of Eastern Washington to the markets of the Sound and Vancouver's Island.

Seattle is built upon the face of rather a steep slope; is pleasant and cheerful-looking, and contains about twelve hundred inhabitants. The Territorial University is located here, and is a fine structure—so situated that it can be seen for a long distance up and down the Sound. Seattle has a great extent of wharfage, which impresses us with the conviction of its business capacities. And, indeed, the harbor swarms with every description of water-craft, from the handsome steamer *Olympia* and the tall three-masted lumber ships, to the little, wheezy tug and graceful "plunger."

On the opposite side of the Sound are Ports Blakely and Freeport; the one a high, round promontory, and the other a long, low neck of land, projecting into the Sound so as to form a small bay with the first. Across the Sound, and nearly abreast of Seattle, is Port Madison, distant twelve miles, also situated on an inlet, so narrow as to compel our steamer to back out—there being no room for "rounding out." All these ports, like Seattle, are great lumbering establishments, and have each a village of from one to three hundred people depending on the mills for employment. Probably one-half the lumbering business of the Sound concentrates within twenty-five or thirty miles of Seattle.

The hay, vegetables, fruits, and provisions generally, that are consumed by the non-agricultural portion of these communities, are furnished by the county of which Seattle is the county-seat.

Port Madison is more handsomely situated than Seattle. It lies on a smooth hill-side, and the residences all have an air of cozy comfort quite prepossessing. One charming feature of the scenery here is the magnificent growth of maple-trees, reserved for ornament and shade. Those trees are as large in proportion to others of their species as are the immense firs of the Sound to theirs: a fact which suggests something with regard to the soil bordering the waters of Washington Territory.

The Port Madison mill is one of the largest on the Pacific Coast. It is 334 feet long by 60 wide; and its machinery is propelled by two engines with eight forty-two-inch flues. Sixty men are employed about the mill, besides the many engaged in logging, rafting, etc. There are shops of every description necessary to a complete establishment, including a foundry and machine-shop. The Company own six vessels for transporting their manufactured lumber; and a steam-tug for towing rafts or vessels, as required.

Port Gamble, thirty miles farther down the Sound, on the west side also, is situated near the head of Tukalet Bay; and is reached by the steamer going fourteen miles out of her course. In its general features Port Gamble is not unlike the other milling establishments, though it is perhaps the most important in respect of the amount of lumber produced. There are two mills at this place, and quite a village of their employees. In passing up this inlet we get a peep into that remarkable arm of the Sound called Hood's Canal, which is between forty and fifty miles in extent, yet whose entrance looks scarcely too wide for the passage of a ship. On this narrow strait is situated Seabeck, another large lumbering establishment; making in all four of these great lumber factories, whose assessed valuation for 1870 was as follows: Port Blakely, $91,-705; Port Madison, $173,191; Puget Mill Company (Port Gamble), $282,327; Washington Mill Company (Seabeck), $128,186. Probably the real value of the milling property in Kitsap County far exceeds the figures which appear on the assessment-roll.

Port Ludlow, seven miles below Port Gamble, and Port Discovery, ten miles south-west of Port Townsend, at the extremity of another of the countless narrow bays by which the Sound is fringed, are the most northern of the milling towns. This latter bay is twelve miles long, and opens into the Sound at its junction with the Strait of San Juan de Fuca.

Port Townsend is situated on the peninsula formed by Port Discovery Bay on one side, and Port Townsend Bay on the other. This peninsula is ten miles long, and about three wide. The shore here is high and abrupt, without trees, and shows a level country beyond. The business portion of the town is located on low ground, only fairly above the reach of the tide, while the residences are nearly all upon the bluff. Though the water is deep, and there is plenty of room in this harbor, it is too much exposed to winds from all points of the compass to be a good one, or to compare favorably with very many others on the Sound.

This is the port of entry for this district; and we make quite a lengthened call, having an opportunity to take a critical look at the group of ladies and gentlemen who have come down to the wharf to give us

17

"hail and farewell." And it must be testified that these people over on the Sound are by no means in a state of darkness or depression, notwithstanding their isolation ; but wide-awake, intelligent, courteous, and modish. The population of Port Townsend is about five hundred.

Two miles and a half south-west of the town is the site of a United States military post, now abandoned. The prospect from this high bluff is remarkably fine. To the north-east, and nearly on the 49th parallel, is Mount Baker, with its ragged, double peak fretting the heavens. Far to the south-east is Mount Rainier, the most beautiful peak of the northern Andes; on the west, the Olympian Range ; on the east, Whidby's Island, spread out like a garden ; and across the straits, San Juan and Vancouver's islands dimly visible.

Leaving Port Townsend, we soon get a view of New Dungeness Light-house on the south, and the San Juan Archipelago on the north—the latter of which recalls the dispute about boundaries : the United States claiming that the English channel should be to the westward of the principal island, and Great Britain that it should be to the eastward. Looking back to the east, across the straits, we see still our mighty snow-peaks towering over a blue mountain-range, with an archipelago of islands intervening. On the southern view, the Olympian Ranges seem to bathe their feet in the waters of the strait, surpassingly beautiful in outline, delicately colored, tipped and rimmed with silvery lines and crests of snow—a marvel of aerial effect—a poet's dream—a vision of the air. Turning from this exquisite sublimity, we see on the north the rocky, but picturesque shores of Vancouver's Island, belonging to the possessions of Her

Britannic Majesty. Neither high nor low, but rising handsomely out of the water; indented with numerous coves, bays, and arms of the sea; its shore being dotted with trees, rather than heavily wooded, with some handsome villas in sight from the steamer, Vancouver's Island makes a good impression at the moment of approach—has, in fact, one of the handsomest approaches to its principal city of any country fronting on the Pacific.

It is impossible to have seen Victoria, the capital of the British possessions on the Pacific, and not give it a passing notice at least. It is so in our way when traveling on the North-west Coast, as not only not to be avoided, but to seem as one of our own proper belongings. The ocean steamers from San Francisco and Portland, though they no longer make this a point of destination as they did in the times of mining excitement on Frazer's River, still call here. The Sound steamers run direct between Olympia and Victoria. There is a large American element in the place, and its contiguity to American soil very strongly suggests identical interests.

The harbor of Victoria is very small, with a narrow and crooked entrance. The site of the town was not selected with reference to a future metropolis, but only as a supply-station of the Hudson's Bay Company, after their removal from Fort Vancouver on the Columbia River. The beauty of the location probably had its proper weight with the gentlemanly managers of the Hudson's Bay Company. Should the future of the city ever demand it, Esquimault Harbor could be opened into Victoria Harbor by a canal across the peninsula on which the city stands, in which case there would be ample room and depth of water.

The discovery of gold, in 1858, caused the British Government to revoke its grant of exclusive right to trade in the North-west Territories, which for so many years had been held by the Hudson's Bay Company, and to erect a new colony, under the name of British Columbia. Under the stimulant of this act, and the repeated new gold discoveries, Victoria suddenly arose from a trading-post to a handsome city of several thousand inhabitants. But her career was brief, owing to several causes, some of which were local and physical, while others were political and traditional. The physical causes for the reverses at Victoria were the severity of the winters in the richest mining region; the cost of getting there, and of subsistence after getting there—the scant agricultural resources of Vancouver's Island not affording provision for the large population which suddenly poured in upon British Columbia. The political ones were those which usually beset a Crown colony, with an expensive Government to maintain, and none but second-hand representation. This brief history will explain why fully one-third of the houses stand vacant in this beautiful city, and point the meaning of such an advertisement as this, in the morning paper: "Wanted—a small family to occupy a house. Rent free."

We also learn by the colonial papers, that railroad schemes, confederation schemes, and annexation schemes are popular topics in Victoria. While one paper declares the "Canadian Pacific Railway a necessity, a possibility, and a certainty," its opponent doubts these assumptions, and sees plenty of obstacles in the way of the proposed railroad. The impartial observer may admit the feasibility of a railroad from the Gulf of St. Lawrence to the Gulf of Georgia, and its seeming benefit to the British colonies—supposing the British Government willing to furnish the means of building it. But its real benefit would still be dependent upon certain political conditions not yet effected. In the present state of the colony, it can not take advantage of the credit of the home Government, nor even extricate itself from the embarrassments which prevent its doing so. To ameliorate this condition, two plans are proposed: confederation with the Eastern Provinces, or annexation to United States territory. Property owners on the island are generally in favor of annexation. But Government officials, and a class of freshly imported young men who have nothing to lose, are opposed to it. That is about the way it stands. Official integrity, patriotism, and British pride are opposed to annexation; and every other interest is in favor of it.

They have not been able always to afford a line of steamers for themselves, and were compelled to see the American steamers pass on up the Sound, only touching at their wharves. This year, however, a line has been put on between San Francisco and Victoria, carrying Her Majesty's mails, and connecting with the Sound steamers, which run to that port. They are forced, too, to buy a great share of their provisions from American dealers. But, in return, they sell their coal (Nanaimo) to American purchasers. Of the retail trade in miscellaneous articles, American merchants in Victoria control a large proportion. We can not help thinking, if it were annexed, what a glorious city Victoria might become. But remaining as she is, she is too near to American enterprise not to be injured by it; and the same is true of British Columbia all the way to the Lake of the Woods. The Northern

Pacific Railroad will draw to itself the agricultural and mineral wealth of a large extent of territory, whose resources are as ample, as they are at present little understood by Americans.

There is little good land on Vancouver's Island, the largest body of it lying on the east side, toward the Gulf of Georgia. Sheep-raising is carried on to a considerable extent; the fleeces being reported light, but the mutton excellent. Fruit, so far as cultivated, does well. But the wealth of the island lies in the heavy forests of the interior, in its coal, and in its fisheries; to which will ultimately be added gold, copper, and salt.

CHAPTER XXIII.

In proceeding from Victoria to Nanaimo—the coaling-station of the American sea-going steamers—we pass through the Archipelago of San Juan, which lies between the Sound and the Gulf of Georgia, and have an opportunity to be surfeited with the beauty of unbroken solitudes. After passing San Juan Island—on which are garrisons of both English and American soldiers—the grounds of the former beautifully laid out, and shaded with spreading maples, we begin to see island after island, all densely wooded, some with mossy banks, overhung with handsome shrubbery, others with bold, rocky shores, of every form of the picturesque. So limpid are the waters that fish, and even sea-weed, can be discerned in their clear depths. These islands were formerly very thickly populated by the native tribes; and many Indians still live about this part of the Sound. Canoes are often met with; and a contrivance for catching wild fowl is frequently observed, which would probably puzzle the regular sportsman who should stumble upon it. It consists of a tall cross, with a net spread on the arms. At night a fire is lighted, which attracts the fowls flying by night, who rush against the net with such momentum as to occasion their fall, when the Indians gather them up before they recover from the shock. Deer are also taken by means of torches, which the Indians burn near their salt-licks. So dazzled are they by the glare that they stand motionless to be killed. From these facts we may gather that the ideal Indian Hunter is a creation far superior in manly attributes to the real one. But then "his bread-and-butter depends on it," and who can blame him?

Bellingham Bay is sixteen miles long from north to south, and about six miles wide. It is, however, divided by islands, which make the bay proper about six miles in diameter, and of an irregular, circular shape. It is backed by rather high hills, covered with forest, which shelter it from the main-land side; and is protected from the winds which blow up the straits by the numerous islands in front. This bay is the shallowest part of the great archipelago, and has a good bottom for holding, with from seven to twenty fathoms in the central division, while in some other portions of it there are thirty fathoms. Of all the countless safe and convenient harbors on the Sound, Bellingham Bay is esteemed one of the most, if not *the* most important, by reason of its nearness to the straits, its excellent anchorage, and its avoidance of the strong currents, which, with the ebb and flow of the tide, set through the narrower channels of the lower Sound. The tides, it is observed, contrary to rule, are highest by night through the summer, and highest by day during the winter; except at the full and change of the moon, when they have their extreme height at six P.M. in summer, and at six A.M. in winter. The average rise and fall is twelve feet in summer and fourteen in winter.

The handsome blue sandstone used in Portland to build the Custom-house is quarried in a little cove of Bellingham Bay, called Chuckanuts. The rock is beau-

tifully stratified, and splits almost as straight as cedar.
It stands in the quarry almost perpendicular to its stratification, and is split in vast surfaces of nearly smooth
stone. It is quarried by cutting through a wall of the
stone on a level with the wharf, then making up-and-
down cuts, and lateral ones, until the blocks are of a
size small enough to be handled; after which they are
slid down to the floor of the quarry to be dressed, or
shipped in the rough.

Coal was first discovered at Bellingham Bay in 1852,
by a Captain Pattle, in the service of the Hudson's
Bay Company, while looking out for spars. A company attempting to work it failed; but another seam
being discoverd a little farther to the north, at a place
called Seahome, another company was formed to work
this one, and succeeded. It was subsequently sold to
a capitalist of San Francisco, who, after a large outlay,
is making it profitable. Miners call this one of the
most regular coal seams that is known. Its thickness
is fifteen feet, with only two divisions of clay. There
appears to be a coal deposit, very little interrupted, all
the way from Frazer River in British Columbia, to the
Columbia River, and beyond.

Whatcom, a mile or two south of Seahome, is the
county-seat of Whatcom County, and a place of probable future importance. It was founded in the beginning of the Frazer River gold fever, in 1858, when as
many as ten thousand people were encamped here
waiting transportation to the mines. A mill was
erected for getting out lumber, and wharves built for
the accommodation of the half-dozen steamers, and
numerous vessels used to carry passengers and freight
to this point. But this splendid prospect for Whatcom
was speedily clouded over. The Governor of British

Columbia issued an order that all miners working in his dominion should take out a license in Victoria. The tide was then turned to Victoria, and thus that city obtained its first great, and as it proved, transient prosperity.

At the mouth of the Lumni, a small river flowing into Bellingham Bay, is the reservation of the Lumnis, a hunting and fishing tribe. Of the eight reservations of Washington Territory, the largest is at Tulalip, on the Sound, east of Whidby's Island. The others are on the Yakima, Chehalis, and Puyallup rivers, east of the Sound; and on the Skokomish River, emptying into Hood's Canal, the Quinaielt River, emptying into Gray's Harbor, and on Neah Bay. All these reservations occupy about two hundred thousand acres of excellent land. This one, on the Lumni, is small, not containing more than twenty thousand acres; but is valuable for its fertility, and the amount of fine timber upon it. The Lumni Indians are very contented, and live comfortably. There are fifty or more board dwellings of a substantial character in their town, which they keep with considerable neatness and order. They are Catholics in religion, observing the forms taught them quite zealously, and seldom neglecting their morning and evening prayers. Generally speaking, the Indians of Washington are better looking, more dignified, and decently dressed, than in Oregon. It is to be hoped that the Government will so deal with them as to save some of these tribes from the degradation and ruin which have nearly exterminated the Oregon Indians. We quote here from the journal of a gentleman who traveled up the Lumni River, on an expedition to Mount Baker:

"Our journey was henceforth up the Lumni, into

the bosom of the forest. Its banks are adorned with several species of willow, alder, the crab-apple, grasses, English clover, the daisy, the cockspur-thorn, the sweet-brier, the wild-rose, and the beautiful festoons of the wild pea. There is plenty of open land, and half a mile up we observe the telegraph wires crossing the river—a silent prophecy of their speedy settlement. . . . Our canoe was propelled against the stream at times by paddles, and at times by poles, and made about three miles an hour. This was slow progress, but we did not regret it, as the scenery became surpassingly beautiful. There were long rows of cottonwood-trees, which, at first sight, reminded one of the English elm. The cottonwood is sometimes called the balsam-poplar. In the spring, when the buds are breaking, the air is filled with the scent of it. Then there would be successive rows of pines in serried ranks, mingled with the cedar and broad-leafed maple, relieved by the gorgeous crimson and Indian-yellow tints of the vine-maple and hazel. The scene would then change : there would be next long reaches of alder and willow, indicating good bottom-lands. Now and then the stately ranks of pines would be broken by some tall fir gracefully leaning forward with its arms, and sweeping the stream like some disheveled beauty. Conspicuous among the arborage is the Menzies spruce (*Abies Menzii*), so called from its discoverer, the surgeon of Vancouver's expedition. Its feathery foliage hangs down in delicate clusters, like lace upon a lady's jeweled arm. Coleridge has said the birch was 'Lady of the Woods,' and we certainly rank the Menzies spruce as the 'Queen of the Forest.'"

From this extract, it will be seen that the same kinds

of trees which have been described as belonging to the rivers and forests of Oregon, extend to the northern limits of Washington; and also that the scenery of this northern latitude loses nothing from being so near the 49th parallel.

Whatcom County, although not yet devoted to grain-raising, is found to produce large crops of wheat on the bottom-lands. It is, however, celebrated for its vegetables, the yield and excellence of all roots, such as onions, potatoes, and turnips, being prodigious. Potatoes are shipped from here to San Francisco. Apples, pears, cherries, plums, and all kinds of berries come to perfection in the region of Bellingham Bay.

Returning down the Sound, the steamer calls at Coupeville, on Whidby's Island — "the garden of Washington Territory." This island is about fifty miles long, and of very unequal width, not being over ten miles at any place. It is almost cut in two by Penn's Cove, one of the long bays common to this region. It is pierced with these inlets in every direction, and receives from them a greater variety of scenery, and greater number of beautiful locations for building, than any equal amount of territory in America. It contains a population of 550, and has about seven thousand acres under cultivation. The excellence of the soil, beauty of the scenery, and mildness of the climate, have given to Whidby's Island a wide reputation. The land is much of it prairie, equally well adapted to farming or grazing. The views which may be obtained from its most elevated portions are remarkably fine, having water, forests, and mountains on every hand. The average mean temperature of the island is forty-eight degrees. Well might so favored a spot be called the garden of the Territory. All that

is true of the most favored portions of Washington,
with regard to grains, vegetables, fruits, and flowers,
apply most especially to Island County, and to Whidby's Island in particular.

We might go on endlessly, describing the many
islands that dot the Sound, and the lovely little bays,
with their small rivers and fertile valleys opening into
them ; but it would be only to repeat the same general
features : agreeable scenery, mild climate, prolific soil,
with a recapitulation of natural resources—animal,
vegetable, and mineral—that are nowhere lacking in
all this immense region of Puget Sound.

CHAPTER XXIV.

THE WASHINGTON COAST.

In order to visit the two most important points on the coast of Washington Territory, we will return to the Columbia River and Astoria. Crossing over by the mail-steamer to Baker's Bay, we find a stage awaiting us by which we are to be conveyed to Oysterville on Shoalwater Bay. The entrance to this bay is twenty-seven miles north of the Columbia, though it extends down to within three or four miles of Baker's Bay, leaving a long strip of land, from one to one and a half miles in width, between itself and the ocean. It is on this long peninsula that Oysterville is situated, and the drive to it is along the beach nearly the whole distance. Of a fine summer's day the excursion is an exhilarating one. The town is upon the inside of the peninsula, and fronts the bay and the main-land opposite. Its distance below the entrance to the bay is about eight miles.

Although the county-seat of Pacific County, and, like Port Townsend, a place for the receipt of customs, it is but a small village, and depends on the oyster trade for its chief support. The annual shipment of oysters to San Francisco is estimated at forty thousand baskets. Quite a number of visitors may be found here in the summer, who come to the coast to escape the heat of the valleys in the months of July and August. The drive on the beach, and the privi-

lege of boating on the bay, are about the only amusements.

Shoalwater Bay is about twenty-five miles long by
from four to seven wide. As its name indicates, the bay
has many shoals, but with numerous deep channels,
which make it easily navigable. There is plenty of
water on the bar—mean low-water being eighteen feet,
and mean high-water twenty-four. There is a light-
house on Toke's Point, the extreme north-west point
of Cape Shoalwater, at the north side of the entrance.
The light is of the fourth order, fixed and varied by a
flash. There are many fine sites for building, both on
the peninsula and on the main-land opposite.

Several rivers empty into Shoalwater Bay—North,
Cedar, Willopah, Palix, Nema, Nacelle, and Bear. Of
these, the Willopah is most important. Its whole
length is not more than forty miles, yet it is navigable
for vessels of twelve feet draught for a distance of fifteen
miles from its entrance; and its valley contains a large
amount of the richest land in the Territory. Next to
the Willopah is the Nacelle. A portion of the Nacelle
Valley is prairie, and the remainder covered with cottonwood. This country is rapidly settling up, and is
represented as being very handsome, with a soil of
rich, black loam. The valley, it is thought, affords
room enough for about one hundred farms. There is a
large amount of Government land in these small valleys, of the best character, on which colonies of farmers might find excellent farms, at Government price.

Fourteen miles north of Shoalwater Bay is a smaller,
but more important one, called Gray's Harbor, covering an area of about eighty square miles. The country between these two bays is a narrow strip of sandy
prairie near the sea, and back of it small lakes, and

cranberry marshes; in all respects resembling the plains south of the Columbia River. Both have evidently been formed by the sand brought down by the Columbia, and other rivers, and moved by wave and wind into its present position.

The entrance to Gray's Harbor is about three-fourths of a mile in width, with twenty-one feet on the bar at mean low-water, and thirty-one feet at mean high-water. It is considered a good and safe harbor. That which makes the importance of Gray's Harbor, is the fact that it lies only about sixty miles · directly west of the head of the Sound; and that it can easily be brought into connection with the Northern Pacific Railroad by means of the Chehalis River, which empties into it. This river, as perhaps the reader will remember, we crossed twice in going from the Columbia to the Sound. At the first crossing is the little village of Claquato, in the valley of the Chehalis, near the junction of the Newaukum with the latter river. Either at this point, or a few miles to the east, in the Newaukum Valley, there must be a station on the Northern Pacific Railroad, the distance from which to Olympia is thirty-seven miles. The Chehalis River being navigable for light-draught steamers for a distance of sixty miles, in a meandering course, leaves but a short distance to be overcome in reaching the railroad, which would give communication with either the Sound or the Columbia River. It was this fact which induced a company to buy up a large tract of land on Gray's Harbor only a few months ago; and the probabilities are that they will make it one of the most important harbors on the whole coast.

Tide-water extends twenty miles up the Chehalis River, and also into some of its tributaries, making

them also navigable. The valley of the Chehalis is
the most extensive of any in Western Washington.
Nor is its beauty or fertility exceeded by any others.
It belongs to the coast only through the circumstances
we have named, the greater portion of its extent being
east of the Coast Mountains. Gray's Harbor is in
Chehalis County, the shire-town of which is Monte-
sano. The population of both Pacific and Chehalis
counties will not amount to more than twelve hundred.

North of Gray's Harbor, until we come to the Straits
of Juan de Fuca, the coast is unsettled. There is,
however, a large amount of fine, level country between
the sea and the Coast Range—'a much greater extent
than anywhere on the Oregon coast. It would also
appear from the last census returns that grain-raising
is carried on no more extensively in the coast counties
of Washington than Oregon. Chehalis County contains 77 farms, on which were raised 785 tons of hay,
3,345 bushels of wheat, 4,235 bushels of oats, 475
bushels of barley. Pacific County contains 56 farms,
on which were raised 384 tons of hay, 1,100 bushels
of wheat, 1,586 bushels of oats, 30 bushels of barley.
Taking into consideration the isolation of these farms,
and that probably they do not depend on grain-raising
for the profit of their farms, the showing is very good.
Pacific County returns 58 horses, 2 mules, 447 cows,
94 oxen, 389 young cattle, 981 sheep, 144 hogs. The
proportion of horses to the number of farms reveals
the fact that but little farm-work is done which requires horse-power to do; and the amount of stock,
that cows, sheep, and beef-cattle are more profitable
than farm crops. But as a great many of the inhabitants of these coast counties scarcely farm at all, doing
just enough farm labor to supply their families with

18

provisions, which they eke out by hunting and fishing, the returns of the census, always below the actual figures, give an idea of the productiveness of the country by no means discouraging. And again, most of the settlers prefer to take up the rich alluvial bottom-lands, which, before being plowed and sown, have first to be cleared ; consequently, their first grain-fields are small.

From what we have seen, we can safely assert that every part of Western Washington can be made not only self-supporting, agriculturally, but something more. All the grains, hardy fruits, and vegetables will grow luxuriantly, and ripen well, in almost any part of it where settlements can be made. But the great wheat, corn, and sorghum region lies east of the Cascade Mountains, in the Walla Walla, Yakima, and other valleys. Peaches and grapes, too, will be generally raised east of the mountains. Beef, mutton, and wool can be produced both east and west of the mountains, of excellent quality. Yet Eastern Washington will excel in the production of these articles ; while the coast country will furnish the dairy products. So happily have the climate and productions of Oregon and Washington been arranged, that almost every luxury the world's markets afford can be obtained within their own borders.

CHAPTER XXV.

"WASHINGTON TERRITORY contains an approximate area of sixty-eight thousand square miles, or 43,520,-000 acres; of this area, about 20,000,000 acres are prairie, and about the same quantity of timber, the remainder mountains. It is estimated that about 5,000,000 acres of the timbered lands are susceptible of cultivation, the remainder comparatively worthless after the timber is removed. A little over one-third of the entire area is adapted to the pursuits of grazing and agriculture.

"The Cascade Range divides the Territory into two unequal parts—eastern and western—differing widely in topography, soil, climate, and productions: the western portion being densely timbered with fir, cedar, oak, etc., with an occasional small prairie, soil varied, river-bottoms sandy mold, with clay sub-soil; high prairies are gravelly or light sand. There are exceptions, however, of rich soil prairies, particularly in Lewis County, which ranks as the best agricultural county in western Washington Territory. Other counties, however, have excellent agricultural land, but they are mostly timbered.

"The principal productions of grain are wheat, oats, barley, and rye; of fruits, apples, pears, peaches, and plums. Small fruits excel. Strawberries, blackberries, huckleberries, and numerous other berries, indigenous.   Salmon and other fish of commercial value are found in the Columbia River and Puget Sound ; while the mountain streams abound in trout, and the woods in game.   The climate of the western portion of the Territory is mild, humid, and remarkably healthy.   The mean temperature of the mouth of the Columbia and of Puget Sound varies but two degrees ; though the climate of the Sound is much more agreeable, by reason of the absence of the strong winds which in summer prevail along the coast.

"Eastern Washington may be described as a vast rolling plain, traversed in all directions by rivers and creeks, the principal of which is the Columbia, having for its tributaries in this Territory the Snake, Spokane, Walla Walla, Winachee, Okinakane, Yakima, and Klickitat, with many of minor importance.   Into the principal of these—Snake River—empties the Pélouze, Clearwater, Tucanon, and other minor streams.   The soil is uniform, and a change is the exception and not the rule ; being a rich, sandy loam, producing a thick, heavy mat of bunch-grass.   On all the streams there is more or less timber ; but the mountains have to supply lumber, and rails for fencing."

Prolific in all her productions, her principal grains are wheat, corn, oats, barley, rye, and buckwheat. Her fruits are apples, pears, peaches, plums, cherries, grapes, and small fruits of all kinds.   Cattle remain fat the year round on the bunch-grass.   The streams abound in fish.   The climate is one which for salubrity may challenge the world ; with a dry air and clear sky the greater portion of the year ; the mean temperature varying from that of the Sound but three degrees, the greatest difference being in summer, when it is eleven degrees warmer in Walla Walla than at Steilacoom, on the Sound. The comparative temperature of Eastern and Western Washington is as follows :

	Spring.	Summer.	Autumn.	Winter.	Mean.
Steilacoom	49.2	62.9	51.7	39.5	50.8
Walla Walla	51.9	73.1	53.6	41.1	53.2

Manufactures—except of lumber, flour, a few woolen goods, and a small amount of leather—are almost entirely undeveloped.

Ship-building on the Sound is carried on to a considerable extent; but has not increased in the last two years, owing to a dullness in the lumber trade in San Francisco, and consequent cheapening of freights by sailing-vessels. In the year 1869, eighteen vessels, of all descriptions, including five steamers, were built on the Sound; but the following two years witnessed a great falling off in the business of ship-building and lumber manufacturing. A large, one-thousand-ton ship was built last year at Port Madison; and a steamer this year at Seattle. Ultimately this must become the great business of the Sound.

The following is the statement of Hon. M. S. Drew, Collector of Customs for Puget Sound District, for the year ending June 30, 1870 :

Value of goods imported from foreign countries.....$33,105 00
Amount of duties collected........................ 41,326 00

EXPORTS OF DOMESTIC PRODUCE.

Value exported in American vessels................$291,010 00
Value exported in foreign vessels 149,905 00

Total exports............................$440,915 00

Live animals of all kinds......................... $43,713 00
Lumber of all kinds............................... 266,288 00
All other articles................................ 130,914 00

$440,915 00

TONNAGE BELONGING TO THE DISTRICT.

	Tons.
62 sailing vessels	13,711.09
19 steamers	2,015.87
8 scows and barges	140.77
Total tonnage	15,867.73

Vessels cleared during the year: American vessels for foreign countries—115 steamers, 4 ships, 13 barks, 2 brigs, 13 schooners, 2 sloops: total number of vessels, 149; number of tons, 55,-606.25; number of men, 2,105.

Foreign vessels for foreign countries: 6 steamers, 16 ships, 6 barks, 3 sloops: total, 31; number of tons, 19,227.42; crews, 456.

American vessels coastwise: 29 steamers, 11 ships, 18 barks, 1 brig, and 9 schooners: total, 68; number of tons, 31,779.74; crews, 1,092. Total number of vessels cleared, 248; total number of tons, 106,613.41; crews, 3,653.

Vessels entered during the year: American vessels from foreign countries—95 steamers, 1 ship, 10 barks, 1 brig, 18 schooners, and 22 sloops: total, 147; number of tons, 39,840.06; crews, 1,852.

Foreign vessels from foreign countries: 6 steamers, 7 ships, and 3 sloops: total, 16; number of tons, 5,366.57; crews, 62.

American vessels coastwise: 39 steamers, 18 ships, 43 barks, 3 brigs, and 6 schooners: total, 109; number of tons, 55,561.18; crews, 1,853. Total number of American vessels entered, 272; total number of tons, 100,767.81; total number of crews, 3,502.

In the coasting trade belonging to other ports there are eighteen vessels, viz., 1 ship, 12 barks, 1 brig, and 4 schooners: total, 18; number of tons, 7,761.25.

The value of the shipments coastwise can not be obtained from any other source than the mills from which the lumber is shipped, as vessels do not clear from this port unless sailing under a register.

The year's shipment coastwise is estimated at three million dollars, being an increase over the preceding year of nearly three hundred thousand dollars.

Imports coastwise can not be ascertained, as the vessels are not obliged to report at the Custom-house except in certain cases.

The total population of Washington Territory is 23,-995. Of this number 6,699 are in Eastern Washington, and the remainder chiefly in the vicinity of the lower Columbia River and the Sound.

Educational and religious institutions are as far advanced and well supported as it would be possible for them to be in a country with so scattered a population. Society, all over the Territory, is rather above the average, promising a good foundation for the future moral and intellectual culture of the State of Washington.

CHAPTER XXVI.

CLIMATE OF OREGON AND WASHINGTON.

THE physical geography of Oregon and Washington is unique, and gives a great variety of climates. Approaching from the Pacific, we find, first, a narrow skirting of coast, from one to six miles in width. Back of this rises the Coast Range of mountains, from three to five thousand feet high. Beyond this range are fine, level prairies, extending for from forty to sixty miles eastward. Beyond these prairies rises again the Cascade Range, from five to eight thousand feet in height, and having to the east of them high, rolling prairies, extending to the base of the Blue Mountains, which trend south - westwardly, leaving plains and small valleys, to the east, between themselves and the Snake River, which forms the eastern boundary of Oregon and a portion of Washington

These differences in altitude would, of themselves, produce differences in temperature. But the great reason why the change is so great from the coast to the Snake River lies in the arrangement of the mountain - ranges ; and in the fact that the north - west shore of the American continent is washed by a warm current from the China seas. The effect of this current is such that places in the same latitude on the Atlantic and Pacific coasts are several degrees — sometimes twenty degrees — warmer on the latter coast, than on the former. This gives a temperature at which great evaporation is carried on. The moisture thus charged upon the atmosphere by day, is precipitated during the cooler hours of night in fog, mist, or rain.

In summer, the prevailing wind of the coast is from the north-west, thus following the general direction of the shore-line. It naturally carries the sea-vapor inland; but the first obstacle encountered by these masses of vapor is a range of mountains high enough to cause, by their altitude and consequent lower temperature, the precipitation of a large amount of moisture upon this seaward slope. Still, a considerable portion of moisture is carried over this first range, and through the gaps in the mountains, and falls in rain or mist upon the level prairie country beyond. Not so, however, with the second, or Cascade Range. These mountains, by their height, intercept the sea-fog completely; and while great masses of vapor overhang their western slopes, on their eastern foot-hills and the rolling prairies beyond not a drop of dew has fallen. This is the explanation of the difference in climate, as regards dryness and moisture, between Eastern and Western Oregon and Washington. All other differences depend on altitude and local circumstances.

Notwithstanding the great amount of moisture precipitated upon the country west of the Cascades, the general climate may be said to be drier than on the Atlantic Coast. The atmosphere does not seem to hold moisture, and even in rainy weather its drying qualities are remarkable. Taken altogether, the stormy days in this part of Oregon and Washington are not more numerous than in the Atlantic States; but the *rainy* days are, because all the storms here are rain, with rare exceptions. The autumn rains commence,

usually, in November—sometimes not till December—
and the wet season continues until April, or possibly
till May; not without interruptions, however, sometimes of a month, in midwinter, of bright weather.
About the middle of June the Columbia River is high,
and during the flood there are generally frequent flying
showers. After the flood is abated, there is seldom
any rain until September, when showers commence
again, and prove very welcome, after the long, warm,
but wholly delightful summer. The annual rain-fall
of the Wallamet Valley ranges from thirty-five to
fifty inches. In the Umpqua and Rogue River valleys
it is less; and at the mouth of the Columbia, and
along the coast, both north and south, it is more. At
Steilacoom, on the Sound, it is fifty-three inches.
From a weather-record of over ten years, kept at
Portland, the lowest point in the Wallamet Valley,
we clip the following summing up:

	Pleasant.	Rainy.	Variable.	Snowy.
1858—9 months	180	48	43	4
1859	228	73	47	17
1860	232	72	57	5
1861	224	70	61	10
1862	250	47	52	16
1863	220	82	55	8
1864	252	60	47	7
1865	227	65	63	10
1866	230	73	59	3
1867	244	65	49	7
1868	272	30	55	9
	2559	685	588	96

"Sixty-five per cent. of the above days were without rain or
snow.

"NOTES.—Ice formed December 2d, 1858. In 1859, ponds were
frozen over at times till March 1st—ice never over two inches
thick; very little cold weather in December, 1859; no ice to
speak of.

"January 24th, 1860, the ground froze for the first time this

winter—first ice January 26th.   Ice and frost all gone February
1st.   I planted potatoes February 6th; on the 17th planted
onion-sets and onion-seeds; April 26th, planted corn.

"January 2d, 1862, Columbia River frozen so that the ocean
steamer could not run; thermometer sixteen degrees below freez-
ing-point.   January 8th, snow a foot deep; excellent sleighing.
On the 17th, Wallamet frozen hard enough to cross on foot.   On
24th, ice gone out of Wallamet River.   March 10th, snow all
disappeared.

"January 7th, 1868, Columbia River closed with ice.   On the
11th Wallamet closed over so as to stop the steamers running to
Oregon City, until the 28th.   No rain fell from the 1st of July to
September 3d—63 days—and then none again till October 23d."

The librarian of the Portland Library Association
furnishes us with the following, for the year 1870 :

Months.	Mean Barom.	Snow.	Rain.	Highest Temp.	Lowest Temp.	Mean Temp.	Rainy Days.
January	.....	.50	4.86	63°	21°	42°	16
February ...	29.71	....	4.30	58°	32°	42°	18
March......	29.83	9.50	4.30	63°	18°	43°	14
April.......	30.01	....	4.30	80°	41°	54°	12
May........	30.03	....	1.95	87°	45°	59°	7
June	30.02	....	1.95	95°	55°	66°	4
July........	30.02	....	.20	102°	61°	73°	1
August	29.95	....	.20	97°	57°	72°	1
September ..	30.01	....	.45	91°	53°	68°	3
October.....	30.15	....	.55	77°	36°	54°	2
November...	30.09	....	6.05	66°	37°	48°	7
December...	30.12	....	4.40	61°	18°	38°	12

"Number of rainy days, 97; number of snowy days, 4; total
rain-fall, 33.50 inches—equal to 2 feet 9 5-10 inches: total
snow-fall, 10 inches."

The mean annual temperature at Corvallis, eighty
miles south of Portland, is fifty-three degrees.   At the
mouth of the Columbia, it is fifty-two degrees two
minutes; and at Steilacoom, fifty degrees eight minutes.   Mean winter temperature of the Sound, forty-
one degrees; mean summer temperature, sixty-two
degrees.

East of the Cascades, the arrangement of the seasons is somewhat different. There is much less rain, which comes in showers, rather than in a steady fall; and is confined to the months between September and June. Occasionally, snow falls to the depth of a few inches; and in some winters has remained on the ground a number of weeks. The heat of summer and the cold of winter are each more extreme, but not at their highest or lowest degrees so trying as the same amount of heat or cold would be in a moister atmosphere. The autumn months in this portion of the county are most delightful, with the thermometer ranging from fifty-five degrees to seventy. The phenomenon of the plains is the periodical warm wind which comes from the south—perhaps from the California valleys—and blows over their whole extent, up to the forty-ninth parallel. The following is the mean temperature for the different seasons, as well as for the year:

	Spring.	Summer.	Autumn.	Winter.	Mean,
Walla Walla, W. T....	51.9	73.1	53.6	41.1	53.2
Dalles, O.............	53.0	70.3	52.21	35.6	52.8

A country like Eastern Oregon and Washington, without marshes or any local causes of miasma; with a clear, dry atmosphere, warmed by the sun, and cooled by the vicinity of snow mountains, could never be unhealthy. Only the most reckless disregard of health can occasion disease in a climate so naturally free from miasmatic poison as this. Western Oregon and Washington may be said to be equally healthful; with this difference, that during the rainy season those who are already invalids are liable to greater depression of the vital powers by reason of the continuous wet weather. Even a well person, of a peculiarly sensitive temperament, may suffer from the same cause. But the age attained by the original settlers of the country, shows well for the general salubrity of the climate of every portion of it. The rate of mortality for Oregon in 1860 was one in 173—the lowest death-rate in any of the States. In 1870 it was one in 155, or about the same as Minnesota.

One peculiarity of the climate of every part of Oregon and Washington, is, the comparative coolness of the nights. No matter how warm the days may have been, the nights always bring refreshing sleep, usually under a pair of blankets, even in summer. Nor does the heat, however great, have that fatal effect which it does in the Atlantic States. Not only men, but cattle and horses, can endure to labor without exhaustion in the hottest days of summer; and sun-strokes are of very rare occurrence.

CHAPTER XXVII.

In Oregon, the forests are found almost exclusively on the mountains. Along the margins of streams there is usually a belt of timber a quarter of a mile in breadth. On the Columbia, this belt, even on the low grounds, is wider; but as there is a range of highlands of considerable elevation extending from the mouth of this river to and beyond its passage through the Cascade Mountains, with only occasional depressions, there is a great body of timber within reach of tide-water.

The base of the Coast Mountains on the west comes within two to six miles of the sea, and frequent spurs reach quite to the beach, forming high promontories. From the coast to the eastern base of the Coast Mountains, is a distance of from twenty to thirty miles. Allowing for the margin of level land toward the sea, and for openings among the foot-hills on the eastern side, here is an immense body of forest lands extending the whole length of the State, from north to south.

Again, the Cascade Range has a base from east to west of about forty miles, including the foot-hills. All the western side of this range is densely wooded, making another great supply of timber. The eastern side, having an entirely different climate, does not support the same heavy growth of trees.

These forests furnish a most interesting study to the botanist. Beginning our observations on the coast, we find that near the sea we have, for the characteristic tree, the black spruce (*A. Menziesii*). It grows to a diameter of eight feet, and to a considerable height, though not the tallest of the spruces. Its branches commence about thirty feet from the ground, growing densely; while its leaves, unlike the other species, grow all round the twig. The foliage is a dark-green, with a bluish cast. The bark is reddish, and scaly; and the cones, which grow near the ends of the branches, are about two inches in length, and purplish in color. In appearance, it resembles the Norway spruce. It loves a moist climate and soil; growing on brackish marshes, and inundated islands

The Oregon cedar (*Thuya Gigantea*) grows very abundantly near the coast. This tree attains to a very great size, being often from twelve to fifteen feet in diameter; but is not so high as the spruce. The branches commence about twenty feet from the ground. Above this the wood is exceedingly knotty; but the lumber obtained from the clear portion of the trunk is highly valued for finishing work, in buildings, as it is light and soft, and does not shrink or swell like spruce lumber. For shingles and rails it is also valuable, from its durability.

The Indians make canoes of the cedar, nearly as light and elegant as the famous birch canoes of more northern tribes. Formerly they built houses of planks split out of cedar, with no better instrument than a stone axe and wedge. An axeman can split enough in two or three days to build himself a cabin. This tree is nearly allied to the *arbor vitæ*, which it resembles in foliage, having its leaves in flat-sprays, that look as if they had been pressed. On the under side of the spray is a cluster of small cones. The bark is thin, and peels off in long strips, which are used by the Indians to make matting, and a kind of cloth used for mantles to shed the rain. It is also used by them to roof their houses, make baskets, etc. Altogether, it is the most useful tree of the forest to the native.

Hemlock spruce (*Abies Canadensis*) is next in abundance near the coast. It grows much taller than the cedar, often to one hundred and fifty feet, and has a diameter of from six to eight feet. The color is lighter, and the foliage finer, than that which grows in the Atlantic States; and the appearance of the tree is very graceful and beautiful.

Another tree common to the coast is the Oregon yew (*Taxus Brevifolia*). It is not very abundant; grows to a height of thirty feet, and flourishes best in damp woods and marshy situations. The wood is very tough, and used by the Indians for arrows. When much exposed to the sun, in open places, the foliage takes on a faded, reddish appearance. It bears a small, sweet, coral-red berry, of which the birds are very fond.

A few trees of the red fir (*Abies Douglassii*) occur in the Coast Mountains, but are not common; also, an occasional white spruce (*Abies Taxifolia*), and north of the Columbia small groves of a scrub pine (*P. Contorta*) appear on sandy prairies near the sea-beach. It grows only about forty feet high, and has a diameter of two feet.

Of the broad-leaved, deciduous trees, which grow near the coast, the white maple (*Acer Macrophyllum*) is the most beautiful and useful. It grows and decays rapidly — the mature tree attaining to the height of eighty feet, and a diameter of six feet; then decaying from the centre outward, lets its branches die and fall off, while from the root other new trunks spring up, and attain a considerable size in four or five years. The wood has a beautiful grain, and is valuable for cabinet manufactures, taking a high polish. The foliage is handsome, being very broad, and of a light green. In the spring long *racimes* of yellow flowers give the tree a beautiful and ornamental appearance, which makes it sought for as a shade tree.

The Oregon alder (*Alnus Oregona*) is another cabinet-wood of considerable value. The tree grows to a height of sixty feet, with a diameter of two or three feet. It has a whitish-gray bark, and foliage much resembling the elm. On short stems, near the ends of the branches, are clusters of very small cones, not more than an inch in length. When grown in open places, with sufficient moisture, it is a graceful and beautiful tree.

Two species of poplar are found near the coast—the cottonwood (*Populus Monilifera*) and the balsam-tree (or *P. Augustifolia*). They are found on the borders of streams, and by the side of ponds or springs; but not so abundant near the coast as east of the Coast Mountains.

Along the banks of creeks and rivers grows one kind of willow (*Salix Scouleriana*), about thirty feet in height, and not more than a foot in diameter, with broad, oval leaves; of very little value.

The vine maple (*A. Circinatum*) is more a shrub than a tree, seldom growing more than six to twelve inches thick near the ground; and not more than twelve to twenty, rarely thirty feet in height. It grows in prostrate thickets, in shaded places, twining back and forth, and in every direction. The wood being very tough,

19

it is almost impossible to get through them; and they form one of the most serious obstructions to surveying, or hunting, in the mountains. The leaf is parted in seven dentated points, and is of a light green. These bushes make a handsome thicket at any time from early spring to late autumn—being ornamented with small red flowers in spring, and with brilliant scarlet leaves in autumn.

Another shrubby tree, which makes dense thickets in low or overflowed lands, is the Oregon crab-apple (*Pirus Rivularis*). This really pretty tree grows in groves of twenty feet in height, and so closely as with its tough, thorny branches to form impenetrable barriers against any but the smaller animals of the forest. The fruit is small and good-flavored, growing in clusters. The tree is a good one to graft upon, being hardy and fine-grained.

Another tree used to graft on is the wild cherry (*Cerasus Mollis*), which closely resembles the cultivated kinds, except in its small and bitter fruit. In open places it becomes a branching, handsome shade tree, but in damp ravines sometimes shoots up seventy feet high, having its foliage all near the top.

When we undertake to pierce the woods of the Coast Mountains, we find, in the first place, the ground covered as thickly as they can stand with trees from three to fourteen feet in diameter; and from seventy to three hundred feet in height. Wherever there is room made by decay, or fire, or tempest, springs up another thicker growth, of which the most fortunately located will live, to the exclusion of the others. Every ravine, creek, margin, or springy piece of ground is densely covered with vine-maple, cottonwood, or crab-apple.

As if these were not enough for the soil to support,

every interstice is filled up with shrubs: some tough and woody; others, of the vining and thorny description. Of shrubs, the sallal (*Gaultheria Shallon*) is most abundant. It varies greatly in height, growing seven or eight feet tall near the coast, and only two or three in the forest. The stem is reddish, the leaves glossy, green, and oval, and the flower pink. Its fruit is a berry of which the Indians are very fond, tasting much like summer apple. This shrub is an evergreen.

Three varieties of huckleberries belong to the same range—one an evergreen, having fruit and flowers at the same time. This is the *Vaccinium Ovatum*, with leaves like a myrtle, and a black, rather sweet berry. The second has a very slender stalk, small, deciduous leaves, and small acid berries, of a bright scarlet color. This is *V. Ovalifolium*. The third— *V. Parvifolium*— resembles more the huckleberry of the Eastern States, and bears a rather acid blueberry. In favored localities these are as fine as those varieties which grow in Massachusetts or Michigan. In addition to these is a kind of false huckleberry, bearing no fruit; and a species of barberry, resembling that found in New England.

Of gooseberries there are also three varieties, none of them producing very good fruit. They are *Ribes Laxiflorum, Bracteosum*, and *Lacustre*.

The salmon berry (*Rubus Spectabilis*) is abundant on high banks, and in openings in the forest. It resembles the yellow raspberry.

Of plants that creep on the ground there are several varieties, some of them remarkably pretty. Of wild roses, *spirea*, woodbine, mock-orange, thorn bushes, and other familiar shrubs, there are plenty.

The Devil's Walking-stick (*Echinophanax horridum*)

is a shrub deserving of mention. It grows to the height of six feet, in a single, thorny, green stem, and bears at the top a bunch of broad leaves, resembling those of the white maple. When encountered in dark thickets it is sure to make itself felt, if not seen. Add to all that has gone before, great ferns—from two to fourteen feet in height, with tough stems, and roots far in the ground—and we have the earth pretty much covered from sun and light.

These are the productions, in general, of the most western forests of Oregon. When we try to penetrate such tropical jungles, we wonder that any animals of much size—like the elk, deer, bear, panther, and cougar—get through them. Nor do all these inhabit the thickest portions of the forest, but the elk, deer, and bear keep near the occasional small prairies which occur in the mountains, and about the edges of clearings among the foot-hills, except when driven by fear to hide in the dark recesses of the woods. In the fall of the year, when the acorn crop is good in the valley between the Coast and Cascade mountains, great numbers of the black bear are killed by the farmers who live near the mountains.

As this region just described is, so is the whole mountain system of Western Oregon and Washington. Along the eastern slope of the Coast Range, around Puget Sound, along the Columbia highlands above a point forty miles from its mouth, and on the western slope of the Cascades, the same luxuriance of growth prevails. Indeed, nearly all the trees enumerated—the black spruce and scrub pine are exceptions—belong equally to the more eastern region. And the same of the shrubs.

But in this more eastern portion grow some trees

that will not flourish in the soil and climate of the coast. Of these the most important is the red fir (*Abies Douglassii.*) Very extensive forests of it inhabit the mountain-sides and Columbia River highlands. It grows to a great height, its branches commencing fifty feet from the ground. The bark is thick, and deeply furrowed; the leaves rather coarse, and the cone is distinguished from other species by having three-pointed bracts between the scales.

The red fir is more used for lumber than any other kind, though it is of a coarse grain and shrinks very much. It is tough and durable, if kept dry. It is a very resinous wood, from which cause large tracts of it are burnt off every year. Yet it keeps fire so badly in the coals, that there is little danger of the cinders carrying fire when buildings constructed of it are burned : it goes out before it alights.

The yellow fir (*A. Grandis*) is also a tree which does not like sea-air, and is very valuable for lumber. It is distinguishable at a distance by its superior height, often over three hundred feet, and by the short branches of the top, which give it a cylindrical shape. It is admirably adapted for masts and spars, being fine-grained, tough, and elastic. The best of lumber is made from this fir, and large quantities of it exported from the Columbia River. The bark of the yellow fir is smoother, not so deeply furrowed as the red, and the oval cone is destitute of bracts.

Of foliaceous trees not found on the coast, is the oak (*Quercus Garryana*), which does not attain a very great size, not growing more than fifty feet high, except in rich, alluvial lands, where they attain fine dimensions. Another and smaller scrub-oak (*Quercus Kellogia*) is common, and the wood is good for axehelves, hoops, and similar uses. The wood of the larger variety is used for making staves, and the bark for tanning.

Of all the trees growing along water-courses, the Oregon ash (*Fraxinus Oregona*) is the most beautiful. In size, it compares closely with the white maple. Its foliage is of a light yellow-green, the leaves being a narrow oval. Like the maple, it has clusters of whitish-yellow flowers, which add greatly to its grace and delicacy of coloring. The wood is fine-grained. and is useful for manufacturing purposes

A little back from the river, yet quite near it, we find the Oregon dogwood (*Cornus Nuttalii*). It is a much handsomer tree than the dogwood of the Atlantic States, making, when in full flower and in favored situations, as fine a display of broad, silvery-white blossoms as the magnolia of the Southern States. As an ornamental tree, it can not be surpassed ; having a fresh charm each season, from the white blossoms of spring, to the pink leaves of late summer, and the scarlet berries of autumn. Its ordinary height is thirty or forty feet, but, in moist ravines and thick woods, it stretches up toward the light until it is seventy feet high.

A poplar not found near the coast, is the American aspen (*P. Tremuloides*). Small groves of this beautiful tree are found about ponds on the high ground, especially where water stands through the rainy season, in hollows which are dry in summer.

A very ornamental wild cherry, peculiar to Oregon —a species of choke-cherry—is found near water-courses. The flowers are arranged in cylindrical *racimes*, of the length of three or four inches, are white, and very fragrant. It flowers early in the spring, at the same time with the service-berry, when the woody thickets along the rivers are gleaming with their snowy sprays.

A broad-leafed evergreen is the arbutus (*A. Menziesii*), commonly called laurel, which is found in the forests of the middle region from Puget Sound, north of the Columbia, to California and Mexico. In Spanish countries it is known as the madrono-tree. The trunk is from one foot to four feet in thickness, and when old is generally twisted. The bark undergoes a change of color annually; the old, dark, mahogany-colored bark scaling off, as the new, bright, cinnamon-colored one replaces it. The leaves are a long oval, of a bright, rich green, and glossy. It flowers in the spring, and bears scarlet berries in autumn — resembling those of the mountain ash. Altogether, it is one of the handsomest of American trees.

On the slopes of the Cascade Mountains is found a beautiful tree — the western *chinquapin* — the flower and fruit resembling the chestnut. Though commonly only a shrub, it here attains a height of thirty feet. This tree is the *Castanea chrysophylla* of Douglas

A very peculiar and ornamental shrub, is the holly-leaved barberry (*Berberis aquifolium*). It has rather a vining stalk, from two to eight feet high, with leaves shaped like holly leaves, but arranged in two rows, on stems of eight or ten inches in length. It is an evergreen, although it seems to cast off some of its foliage in the fall to renew it in the spring. While preparing to fall, the leaves take the most brilliant hues of any in the forest, and shine as if varnished. The fruit is a small cluster of very acid berries, of a dark, bluish purple, about the size of the wild grape, from which it takes its vulgar name of "Oregon grape."

It is very generally removed into gardens for ornament.

In damp places away from the rivers, grows the rose-colored *spirea* (*S. Douglassii*), in close thickets; and is commonly known as hardhack. Near such swamps are others of wild roses of several varieties, all beautiful.

It is almost impossible to give the names of the numerous kinds of trees and shrubs which grow in close proximity in the forests of Oregon. Beginning at the river's brink, we have willows, from the red cornel, whose crimson stems are so beautiful, to the coarse, broad-leafed *C. Pubescens*, ash, cottonwood, and balsam-poplar. On the low ground, roses, crab-apple, buckthorn, wild cherry. A little higher, service-berry, wild cherry again, red-flowering currant, white *spirea*, mock-orange, honeysuckle, low black-berry, raspberry, dogwood, arbutus, barberry, snow-berry, hazel, elder, and alder. Gradually mixing with these, as they leave the line of high-water, begin the various firs, which will not grow with their roots in water. As the forest increases in density, the flowering shrubs become more rare; re-appearing at the first opening.

It would be impossible to exaggerate the beauty of such masses of luxuriant and flowering shrubbery covering the shores of the streams. Even the great walls of basalt, which are frequently exposed along the Columbia, are so overgrown with minute ferns, and vivid-green mosses, and vines, as to be much more beautiful and picturesque than they are forbidding.

In Southern Oregon, the botany of the forest changes perceptibly, some species of the Columbia River disappearing, and others taking their place; the change being accounted for by the greater elevation of the country, and the superior dryness of the climate.

There are several pines common to this more southern region : the sugar pine (*P. Lambertina*), balsam fir (*P. Grandis*), and *P. Contorta*, or twisted pine. Besides these, the manzanita (*Arctorstaphylos, Glauca*) and *Rhododendron maximus* belong to the southern portion of the State.

The game natural to forests and mountains is more abundant also in the southern ranges ; and, from the greater frequency of open or prairie spots in the mountains, much more easily hunted.

The eastern side of the Cascade Range is but thinly timbered, and that with the yellow pine (*P. Ponderosa*), which has a trunk from three to five feet thick, and attains an average height of a hundred feet. The foliage of this pine is scattering, coarse, and longer than that of the Eastern varieties. Few shrubs grow on this slope of the mountains ; and the smooth, grassy terraces have more the appearance of cultivated parks than of natural forest. Along the streams of Eastern Oregon are but few trees; generally the willow, alder, cottonwood, and birch.

Upon the greater elevations of the eastern slope, the western larch (*Larix occidentalis*) appears quite frequently. It is a large tree, tall and slender, with short branches, leaves long and slender, and foliage of a pale, bluish green, light and feathery.

More rarely occurs, in peculiar situations, the silver fir (*Pinus amabilis*), so called from the silvery appearance of the under side of the dark-green foliage. The cones grow erect near the summit of the tree, and are of a size of six inches by two and a half, of a dark-purple color, and rather smooth appearance. The mountain ash also occurs, at rare intervals, in the Cascade Mountains.

Possibly there are other trees and shrubs not mentioned here. Our intention has been to make the reader acquainted with the general features of an Oregon forest ; and if we have not failed in our intention, a comparison of our notes with the trees which compose one will enable him to identify most of them. For their botanical classification, we are indebted to the botanist of the Railroad Exploring Expedition.

Washington Territory contains more large bodies of timber standing on level ground, than Oregon does. An immense extent of fir and cedar forest encircles the whole Sound, and borders all the rivers, besides that which is found on the foot- hills of the Cascade and Coast ranges. It is estimated that three-fourths of Western Washington is covered with forest, a large proportion of which is the finest timber in the world, for size and durability. It is nothing unusual to find a piece of several thousand acres of fir, averaging three and a half feet in diameter at the stump, and standing two hundred feet without a limb—the tops being seventy feet higher. Three hundred feet is not an extraordinary growth in Washington Territory.

It would be impossible, in speaking of the forests of Oregon and Washington, to pass in silence the subject of their commercial importance. It is estimated that the area of forest land in Oregon and Washington covers 65,000 square miles. Not all of this timber is accessible, nor all of it valuable for market, and yet the quantity is immense that is marketable. Puget Sound exports annually from 100,000,000 to 350,000,-000 feet of lumber ; the Columbia River, 20,000,000 ; and the mills along the Oregon coast, about 35,000,000.

The Portland mills manufacture jointly 7,600,000 feet per annum, which is all consumed by the home demand; besides large quantities of planking for streets, which is furnished by the St. Helen mill, on the Columbia River. The amount manufactured all over the country, for home consumption, can hardly be correctly estimated, yet must amount to 75,000,000 more. This seems an enormous sum total for a comparatively new country; and suggests the possibility of sometime exhausting even the great timber supply of Oregon and Washington. The amount manufactured by the mills of the Oregon and California Railroad, for the construction and equipment of that road, can hardly be estimated. One of their mills is capable of cutting 400,000 feet per week.

The kinds of timber adapted to lumbering purposes are known as the red, white, and yellow fir, cedar, hemlock, and, in some localities, pine and larch. The red fir constitutes the great bulk of common lumber; the yellow fir is used where strength and elasticity are required, as in spars of vessels, piles, wharves, bridges, and house-building; and cedar for foundations of houses, fence-posts, and inside finishing of houses.

The cabinet-woods are maple, alder, and arbutus. There is oak for staves, and other purposes; but nothing that answers for wagon-making grows on these mountains. Hemlock becomes valuable as furnishing bark for tanning leather. Ash is used for some mechanical purposes; and makes excellent fire-wood.

The red fir is very resinous, and might be made valuable for its pitch. The quality of Oregon turpentine is superior; but owing to the high freights and high rates of labor on this coast, has not heretofore proven profitable as an export. It is common to find a deposit of dried pitch or resin in the trunks of large fir-trees—especially those that have grown on rocky soil—of one to two inches in thickness, either forming a layer quite round the heart of the tree, or extending for fifty feet up through the tree, in a square "stick."

Trees that have been destroyed by fire have their roots soaked full of black pitch or tar; and even the branches of growing trees drop little globules of clear, white pitch on the ground.   This wood makes excellent charcoal; in the burning of which a great deal of tar might be saved, by providing for its being run off from the pit.   There is also plenty of willow wood for making charcoal, growing on all the bottom-lands.

From the figures given, it will be seen that Washington Territory exports, at the lowest estimate, double the lumber of Oregon.   Puget Sound has unrivaled facilities for that branch of business; and has not much of an agricultural population adjacent to it. Oregon, on the contrary, is chiefly agricultural, yet not without excellent facilities for the manufacture of lumber.   It will take years of lumber-making along the Columbia, and the rivers and bays of the Coast Range, to clear even a mile of water-front in the vicinity of one mill.   This is why fires are suffered to destroy so much fine timber every year : the farmers can not get the heavy growth off the land in any other way.

It follows, of course, that where the supply is so great the price is correspondingly small.   At some country mills, run by water-power, lumber can be obtained for nine dollars per thousand feet.   At Portland, where they are all run by steam, the prices range thus : Street planking, $11 to $12 ; common lumber, $14 to $15 ; siding, $20 to $21 ; flooring, $26 to $28 ;

and miscellaneous dressed lumber, from $20 to $31. The Columbia River mills are not at so much expense in getting logs, and consequently sell at figures somewhat lower.

The cost of manufacturing lumber depends very much upon the location of the mills. Those that are situated on a navigable river, slough, or bay, near a fine tract of fir or cedar, or both, have greatly the advantage; and there are plenty such locations along the Lower Columbia. The St. Helen mill, for instance, gets logs rafted from Scappoose Bay, Lewis River, or the Columbia, from tracts of good timber from three to twelve miles distant. There are splendid bodies of cedar upon Lewis River; also, near the Columbia, two or three miles below St. Helen; and fir all around—on all the tributaries, and on the great river itself. The Rainier and Oak Point mills are in the midst of timber; and so of those on the coast.

The price of logs in the raft is from $3.50 to $5 per M., where it is necessary to purchase them. Labor is worth from $2 to $5 per day, according to the grade; all expenses estimated in gold coin. Timbered lands, conveniently located, are held at from $8 to $15 per acre; but there is plenty of Government land within two or three miles of the Columbia. Mill-owners generally contract for logs to be delivered at their mills, instead of buying up forest lands. The logging business is a very profitable one, being attended with little expense besides the cutting of the logs. There are in all Oregon, as nearly as can be estimated, one hundred and fifty-three saw-mills; thirty-eight being run by steam. Of the mills of Washington Territory there are not so many, but much larger ones, being chiefly engaged in making lumber for export.

CHAPTER XXVIII.

MANY of the flowering shrubs of Oregon and Washington have already been mentioned in the chapter on Forests. One of the first to blossom is the red flowering currant (*Ribes sanguinerum*), which puts forth its flowers before its leaves are fully expanded, like the Judas-tree of the Missouri Valley, which it resembles in color. There appear to be two or three varieties of this species, as the color varies from a pale rose color to a full crimson. The flower is arranged in clusters upon a slender stem, like the green blossoms of the garden currant ; but is much larger, and of a different shape. The bush is highly ornamental when in blossom, and generally introduced into gardens for decoration. It flowers in March. East of the Cascades is a yellow species very similar. Both of these grow near streams, and in the edge of the forest.

Of the *spirea*, there are several species. The waxberry, with its tiny pink flowers and delicate leaves, is found in bottom-lands and on river-banks. In autumn the bottoms of the Columbia furnish thickets of waxberries, which, growing side by side with the wild roses, make a pretty contrast to the crimson capsules of the latter. In higher ground, yet subject to overflow, is found the *Spirea tomantosa*, or hardhack, as it is commonly called, which grows in thickets, and bears a cluster of a purplish-pink color. But the most beautiful of the *spireas* is the kind known as sea-foam (*S. Ariæfolia*), which its great, creamy-white clusters really resemble. This grows along the river-banks, and in the shade of the forest's edge, and blooms in June and July, according to its locality. It sometimes grows to a height of twenty feet, in the shade, though usually about five or six feet high. The stems are very delicate, like all the *spireas*, and bend most gracefully with the weight of the clusters.

Side by side, usually, with the last-named *spirea* is the beautiful mock-orange (*Philadelphus*), with its silvery-white flowers crowding the delicate green leaves out of sight. Throughout Oregon this shrub is called syringa, to which family it does not belong. It is very ornamental, and blooms in June and July.

Of wild roses there are several species, and many varieties, from the dainty little "dime rose," of a pale pink color, to the large and fragrant crimson rose, which grows in overflowed ground. There are always some roses to be found, from June to December. It is usual to find the shrubs here mentioned growing in close proximity; and these, with the flowers of the woodbine (*Lonicera Occidentalis*), and the blossoms of various kinds of wild fruit trees, make a perfect tangle of bloom and sweetness along the river-banks in summer.

We have elsewhere spoken of the dogwood, which is as handsome as a magnolia-tree when in blossom; and of the wild cherries and other fruits whose flowers are sweet and beautiful. The Oregon grape, or holly-leaved barberry, bears a flower that is very ornamental, of a bright yellow color, in clusters a finger long. The leaves of this shrub are also very beautiful, which makes it desirable to cultivate. Its fruit is ripe in August, and is of a bluish-purple, like the damson plum.

In Southern Oregon, the *Rhododendron maximum* is one of the glories of the mountain-tops, with its immense branches of rose-colored flowers. It is occasionally seen in gardens. The buff-colored *Azalea occidentalis* is also confined to the southern and eastern portions of Oregon. It is said that the clematis grows east of the Cascades, but we have not seen it; and also the ilex-leaved mahonia. The wild grape (*Vitis Californica*) is another shrub or vine which is confined to the southern portion of Oregon. In the Rogue River Valley, in October, it is a striking ornament in the landscape; the foliage being turned a rich ruby-red color, and forming clumps upon the ground, or hanging pendent from way-side trees. It does not seem, however, to furnish much fruit.

Of field flowers, there are a great many in all parts of Oregon and Washington; beginning with the early spring to beautify the earth, and kind succeeding kind throughout the summer and autumn. There are, especially near the Columbia, where the soil which covers the rocks is often a thin, black mold, countless varieties and species of very minute flowers, so small frequently as to need a microscope to analyze them successfully; but of lovely shapes and colors. We have found within the range of an acre forty kinds of flowering plants in the month of July, half of them of this minute size.

Of the plants peculiar to the North-west which bear handsome flowers, the Camas family is prominent. The *Camasia esculenta*, or edible camas, of whose roots the Indians make bread, grows about eighteen inches high, and bears at top a bunch of star-shaped flowers, of a beautiful lavender color, with a golden centre. The leaves grow from the root, and are lanceolate.

The places where they are most abundant usually are called "Camas prairies," and they form a feature of Eastern Oregon and Idaho. They are also plentiful in Western Oregon. The flowering season is about the middle of May, near the Lower Columbia. There are several species of the camas, one of which is poisonous, as noticed above.

It would be impossible to any but a thorough botanist to give a complete list of the flowering plants native to this country. We shall, therefore, briefly notice those which are most common, and which we have had an opportunity of observing. Commencing with the spring, we have the purple iris; *mimulus luteus* (yellow); yellow lily (*Golden Erythronium*); white, blue, purple, and yellow lupines; wild pea (*Vicia*); white daisy; California yellow poppy (*Oenothera Biennis*); pink oenothera; verbena; brodiea, belonging to the family of Camas—two varieties—both purple, one of them very beautiful, found near Albany; silene, commonly called a pink, very elegant; tobacco pouch (*Cypripedium*), white, shaded with gray; Indian pink (*Castelia brevifolia*), scarlet, or orange red; shooting star (*Dodecathem media*), several colors; larkspur; flax flower (*Linum*); boys and girls (*Cyno Glissum*), pink and blue on the-same-stem; orange lily (*Lilium Canadense*); red ear-drop (*Delphinium nudicaule*) white dowbell (*Cyclobothra alba*); red columbine; *Lilium Washingtonium*, the great white lily of the Wallamet Valley; pink convolvulus; golden coreopsis; *Phlox, Clarkia;* Anemone; sunflower; golden red; (*Salidago*); aster; dicentra, white and scarlet; *Collomia grandiflora*, salmon color; *Dichelostema congesta* (poison camas), purple; *Hesperoscardum grandiflora*, a white flower, marked with green, very delicate; hossackia bi-color, white and yellow; and others whose names are unknown to us, or which have been forgotten.

Of flowering grasses, and delicate flowering vines that run on the ground in the woods, there are several; but their botanical names are unknown, and they have no common names except in two instances. One of these is a spicy, little, running vine called Oregon tea, common in all woods; and the other is a beautiful myrtle which is found about Puget Sound, in the shade of the giant trees. Of ferns and mosses, there is an endless variety in the woods, and on the rocks of Western Oregon and Washington.

The prevailing colors of wild flowers in Western Oregon are purple, yellow, and white, with a fair proportion of pink or red. In Eastern Oregon, there are still fewer red flowers. Blue flowers are very rare in any portion of this country, as, we believe, they are everywhere. We remember to have seen some lovely blue flowers growing in the sands between Wallula and the first crossing of the Touchet, but we did not get any of them. Buff or salmon color is still rarer, the *Collomia* being the only one we remember seeing. Yet with all the different shades of the common purple, yellow, white, and red, with their differing forms, a great deal of beauty may be expressed; and the prairies of Oregon and Washington, east and west, present a delightful bouquet of tints in the summer months.

Very few flowers of the Wallamet Valley are fragrant; while, on the contrary, very many of those found east of the Cascades are highly perfumed; as they are also in Southern Oregon, where the blue violet, quite scentless near the Columbia, is deliciously fragrant. Of the early spring flowers common to the

Atlantic States and to this country also, are the yellow violet; adder's tongue, or dog-tooth violet; spring beauty, and buttercup. But the spicy wintergreen, with its crimson berry, is unknown; as is also the May apple, and other delights of childhood. A vine resembling the checkerberry is said to grow near the coast, but we never saw it.

The soil and climate of Oregon and Washington is highly favorable to the growth of flowers; and we may find in the gardens here, flowers from almost every clime, growing in more or less perfection. From the plenitude of moisture, they continue to blossom very late in the season; a bouquet of roses, and a dozen other varieties of elegant flowers, being often gathered at Christmas. Frequently, gardening can be resumed in February, which gives a large proportion of the year to the enjoyment of one of the purest and most wholesome of pleasures.

The United States Exploring Expedition collected, in the year 1854–5, three hundred and sixty species of native plants, of which one hundred and fifty are peculiar to the prairies of Oregon and Washington.

CHAPTER XXIX.

NOTWITHSTANDING the thick growth of the forests of Oregon and Washington, the hunter may find sport, with game worthy of his rifle, if he is not afraid of the exertion and foot-service. There are numerous "openings" in the forest, and plenty of wild country in the foot-hills, where game may be found, if the *habitat* of each animal is known.

The most formidable of the Bear family is the grizzly, which inhabits less the thick forests of the north than the manzanita thickets and the scrub-oak coverts of Southern and Eastern Oregon, yet is occasionally found as far north as the Olympian Range in Washington Territory. The color of this bear is a silvery gray, its bulk immense, sometimes weighing two thousand pounds, and its habits herbivorous chiefly, though it will, on sufficient provocation, kill and eat other animals, and even man. It subsists in Southern Oregon upon the berries of the manzanita, of which it is very fond, and will feed upon any berries or fruits within its reach ; occasionally, as a relish, digging up a wasps'-nest for the sake of the honey, not being able, like the black bear, to climb in search of bees'-nests.

In seasons when drought has destroyed its customary food in the mountains of California, it has been known to descend into the valleys and dig up gophers for

food. If it scents fresh venison or beef, it will steal
it if possible, and has been known to take the hunter's
provisions out from under his head while sleeping. In
such a case it is better to pretend to be sound asleep
during the stealing, even if very wide-awake, as most
likely to be the case, for any movement will be certain
to bring down the bear's paw with force upon the
hunter's head—"a consummation most devoutly to
be" avoided.

This trick of the grizzly—striking a man on the head,
or "boxing his ears"—is a dangerous one. It is not at
all rare to find men in the mountains and valleys where
the grizzly ranges, who have had their skulls broken
by the blow of its immense paw. They are much to
be dreaded in a personal encounter, and by no means
easy to kill, unless hit in the vulnerable spot behind
the ear. Those who fancy lion-hunting in the jungles
of Africa might find equally good sport in hunting
grizzlies in California, Oregon, and in some parts of the
Rocky Mountains.

During the summer months they retire to the mountains; but, as the berries ripen, come down to the foot-
hills and river-banks, to feed upon their favorite fruits.
If a cavern is not at hand when winter comes on in
the cold regions they make a bed for themselves in
some thicket; or sometimes dig a hole below the surface, in which they pass the winter sucking their paws.
It would seem that where the winters are as mild as in
the Coast Mountains of California, they do not hiber-
nate, as they are met with all through the winter season, and kill, and are killed, more than ever at that
time, on account of the scarcity of berries.

There are several curious facts in the natural history
of this bear; one of the most singular of which is,

that the period of gestation is entirely unknown, even to the most observant and experienced mountain men. No one has ever killed a female carrying young, at any time of the year, though they are often discovered with their cubs evidently but a few weeks old. Where they hide themselves during this period, or how long it lasts, no hunter has ever been able to observe; though there are men who have spent half their lives in the mountains, and killed in desperate encounter many a grizzly, and at all times of the year, even when hibernating.

The grizzly seems to be "a man of many minds," with regard to attack. Usually, unless in charge of cubs, it quietly avoids a meeting with the hunter; and at times, even seems timid and easily alarmed. But because one grizzly has given you room, you must not depend upon the next one doing the same. It is quite as likely that he will challenge you as you pass; and unless well prepared to take up the glove, you had better "take up" the first tree you come to. It is not a pleasant sight to see one of these monsters on his hind-quarters, with his fore-paws ready for action; and when it comes to running, he can run as fast as you can.

The brown, or cinnamon bear, is also a savage creature, with many of the traits of the grizzly, but inferior in size. They inhabit the same regions with the latter, and also are found in the thick forests of Northern Oregon and Washington.

The black bear is common to every part of these countries, living in the mountains in summer, and visiting the low hills and small valleys, or the banks of rivers, in autumn. When the acorn crop is good in the foot-hills, bears haunt the groves which furnish their favorite food. If they can find a stray porker engaged in foraging, they embrace him a little too tightly for his health—in short, "squeeze the breath out of him"—after which affectionate observance they eat him. But unless exasperated they never attack the human family, and are not regarded as dangerous under ordinary circumstances.

An animal which is ferocious, and not unfrequently met with in the mountains, is the cougar—an animal of the cat species, with a skin something like a leopard's, and a long, ringed tail, but a head with a lion-like breadth. It is variously called the California lion and American panther. We saw one large specimen, which was lying dead by the roadside on the Calapooya Mountain, which measured seven feet from tip to tip. This animal seldom attacks a man, but is very destructive to calves and colts in the vicinity of the mountains, especially in the newly settled parts.

There are three species of the wolf in Oregon and Washington, of which the black is the largest and most ferocious. It stands two and a half or three feet high, and is five to six feet from tip to tip. Such was its destructiveness in the earliest settlement of the country that special means were resorted to for its extermination, until now it is rarely ever met with. It attacks young cattle and colts, as well as the cougar.

The white, or gray wolf, is another enemy to the stock-raiser, though it is satisfied with smaller game than the black wolf, contenting itself with full-grown sheep; and being more powerful than a dog, is a great destroyer of flocks in some localities, and so sagacious that it is very difficult to poison. The coyote, or barking wolf, is also a depredator, taking young pigs and lambs. One of these little animals has the voice of several, and can imitate the barking of a whole pack.

It is almost too contemptible to be considered game, and is given over to strychnine.

There are two or three species of lynx, or wild cat; also troublesome to settlers near the forest, carrying off young pigs, and such small farm stock. When not stealing from the farmer they subsist themselves upon young fawns, hares, squirrels, and game birds. These are numerous in the woods of the Lower Columbia. We have seen several good specimens depending from the limbs of trees, where they had been hung after shooting.

Of foxes, there are the red, silver gray, black, and gray varieties. It is thought that the black fox is a distinct species; as is also the gray, which is smaller. But the silver gray is said by the Indians to be the male of the red species; the female only being of a reddish color. This species, in all its varieties, is very common on the eastern side of the Cascades, near the Columbia; and the smaller gray is most abundant in the Klamath Lake region, in the southern part of Oregon. Their skins, though not as handsome as the silver gray, are still very fine. The gray is the "medicine fox" of the Indians, a meeting with which brings misfortune.

Elk are found both in the Cascade and Coast mountains; but are most abundant in the latter. In summer they keep pretty high up, generally; but when snow falls in the mountains, descend to the plains and river-bottoms. They travel in well-beaten trails, and in large droves, which makes them easy game. When quite wild they show considerable curiosity, stopping to look at the hunter, thus offering a fair shot. When wounded and in close quarters they are formidable antagonists, from their great size, heavy head, and large antlers. The immense size of their antlers would appear to be an obstacle to their escape, when running in the forest; but by throwing back their heads they drop them over their shoulders, so well out of the way as to enable them to pass through the thick woods without difficulty. There are still immense herds of them in the mountains near the mouth of the Columbia, and may be hunted in summer by parties sufficiently hardy for overcoming the obstacles of the forest. But autumn and winter are better seasons for hunting elk, as they then come down to more open ground. Elk-steaks are no rarity in Astoria; and occasionally they are to be met with in the Portland markets. It is estimated that not less than one thousand elks were killed last year in Coos County alone, for the skins only.

Three species of deer are found in Oregon and Washington: the white-tailed, black-tailed, and mule deer. The two first-named species inhabit the country west of the Cascades, the black-tailed being most common. They also inhabit east of the mountains, but have been greatly decimated by the Indians, who kill them wantonly in snowy winters when they can not run. In the mountains along the Lower Columbia and Lower Wallamet they are still very plentiful. Game-laws exist in Oregon for protecting them during a certain season; and still lawless persons are found who kill them without regard to their condition. The mule deer is found only east of the Cascades, and is not common. It seems to be a hybrid between the antelope and black-tailed deer.

The antelope is an inhabitant of Eastern Oregon, and is hunted by the Indians by a "surround"—for though curious enough to stop to look at the hunter, it is very fleet, and soon distances pursuit. Hence the

Indian method of driving them into a corral, by coming down upon a herd from all sides and gradually forcing them into an inclosure made for the purpose—a very unsportsman-like way of taking such delicate game.

Eastern Oregon also furnishes the mountain sheep. In the region of John Day and Des Chutes rivers, they were formerly very numerous. Their flesh is good, though likely to be flavored with whatever they feed most upon. It appears from the testimony of early voyagers to this coast, that the Indians formerly made a kind of cloth from the wool of the mountain sheep, but the process of its manufacture is unknown in Oregon at this period. The fact of the sheep being native to the grassy plains of Eastern Oregon and Washington, furnishes a hint by which wool-growers might profit.

The prairie hare—a large, blue-gray species—is found in Eastern Oregon and Washington; and also on the mountains of Southern Oregon, where it is very common. The flesh is good eating.

Of fur-bearing animals which are hunted for their skins, there is the hair seal in the Columbia River—a very pretty creature, of a bluish-gray color, spotted with white. These seals swim up the river as far as the Cascades, and in high-water make their way up to the Dalles. They are smaller in size than the red seal of the Pacific, and very docile in their dispositions. Instances have occurred of their being domesticated, when they have shown the same attachment to their masters that the dog does, following them also by scent, even into the thick woods, where they have torn themselves fearfully in their efforts to overtake those who had deserted them. The Indians roast and eat them.

The mink, whose fur is so fashionable and valuable, is common to the waters of Oregon and Washington, but most numerous in the lakes and Sound of the latter. It is said that when they inhabit the Sound, they subsist upon shell-fish.

The beaver, which was nearly exterminated in the days of the Hudson's Bay Company's occupancy, is again quite abundant in the streams of all the wooded portion of the country. One of the sights peculiar to the Lower Columbia is the "hunting-boat"—a sort of scow with a house on it—which goes peering into all the creeks and sloughs leading out of the river, after game of this sort. The .skins are taken to Portland, or to some trading-post along the river, and sold or exchanged for goods.

The " California otter" also inhabits the mountain streams in considerable numbers, especially those that come down from the Cascades. The sea-otter is found along the coast, but is becoming rarer; having, it is supposed, left the American for the Asiatic coast.

The pine marten, or American sable, is in considerable numbers along the streams of the Cascade Mountains, and is found clinging to pine-trees on the eastern slopes, in Washington and Oregon. Their skins are quite valuable, though not collected except by the Indians, who prize them for ornament.

Of game birds there are great numbers, as might be conjectured from the nature of the country. The habits and *habitats* of this kind of game are too well known to sportsmen to need remark. We will give the names only of the most common: Mountain quail; valley quail; dusky grouse, ruffled grouse; sharp-tailed grouse, or prairie chicken (found only east of the Cascades); sage-cock (east of the Cascades); curlew (east of the Cascades) ; kildeer plover ; golden plover ; Virginia rail ; English snipe ; red-breasted snipe ; summer duck ; Canada goose ; white-fronted goose ; black brant; mallard duck; canvas-back duck; blue-winged teal ; brown crane ; green-winged teal ; and probably several birds omitted or forgotten.

In autumn, the region of the Lower Columbia is swarming with wild water-fowl. A common recreation among the Portlanders is to charter a small steamer, or in place of it a hunting-boat, to convey a party of gentlemen to the haunts of geese and ducks, among the streams and sloughs about the mouth of the Lower Wallamet, and up into Scappoose Bay. A week's sport, with good living on board their hunting craft, is thought "worth the shot," as unbending both body and mind from the year's routine of business.

When it is remembered that there is the best of sport for the angler in the creeks and rivers of the country, where choice may be made between a seventy-pound salmon at the mouth of the Columbia, and a dainty, speckled trout in almost any tributary, it must be allowed that there is amusement for all varieties of idle people, not to say healthful pastime for invalids, in Oregon and Washington.

There is also here—what can not readily be found in the Atlantic States—a class of men who have made hunting and trapping the business of half their lives ; and who, while they lend their knowledge of the craft to the inexperienced hunter, entertain him with volumes of humorous and exciting personal adventure with every sort of game, from a beaver to a Black-foot Indian. The "River of the West," which chronicles much of this kind of wild life, furnishes an index merely to what the traveler may learn for himself, if he has a few months' leisure to make himself familiar with these men and the scenes of their exploits. The curious traveler may find in Oregon men who were with Sublette, Wyeth, and Bonneville, in the mountains; men who met there Stanley, the painter, Douglas, the botanist, Farnham, the would-be founder of a communist colony; who hunted beavers and Indians with Kit Carson; who laugh at Fremont as an explorer; who served Wilkes on his surveying expedition; who saw Oregon in danger of becoming an independent Government, but whose noble patriotism saved it to the Republic of the United States.

CHAPTER XXX.

No GREAT and general forms of Nature impress themselves upon the memory or imagination more than mountains. The ocean alone rivals them in this respect. Those nations, like the Swiss, who have been born and bred in the shadow of, or even in sight of, cloud - piercing heights, never take kindly to countries of a smoother aspect. In a few generations, Oregon will undoubtedly possess, for this reason, a people distinguished for patriotism.

We have found the Oregon mountains everywhere, west of the summit of the Cascade Range, densely covered with forest, even up to the line of perpetual snow. Considering the impenetrable nature of the forest, this would alone render a passage through them very difficult. Yet this difficulty is not the only one. From the summit of the Cascades to the open country at their base, is a distance of from forty to sixty miles. Not, indeed, a smooth descent, nor a succession of parallel ridges; but a bewildering chaos of mountains, thrown together in such confusion that engineering is dismayed at the task of finding a pass among them.

Yet all the earliest roads into Western Oregon were surveyed by the hardy pioneers, who knew very little of scientific engineering. With a bravery and perseverance most heroic, they struggled with and overcame the obstacles that met them on the last portion of a long and exhausting overland journey. The location of a road from Dalles to Oregon City nearly cost two brave men their lives; but they secured their object. The first train of immigrant wagons which came over this road made from eight to ten miles per day; their forces being occupied most of the time in widening the track—for the pioneers had found it too much labor to open a very broad highway for those who were to follow.

The most skillful driving did not prove skillful enough to guide the staggering oxen through the way provided by the road-makers; and the constant tendency of a forward wheel to run up a tree, on one side or the other, was very trying to the drivers. But if the wagons would run up trees on ascending ground, what was their course when they came to an incline of nearly sixty degrees on the descending side, with a heavy load urging the jaded oxen from behind?

As succeeding trains gradually widened the way, a new difficulty arose. It was better to be stopped by a tree than not to be stopped at all, or to find one's team rushing down the side of a mountain, like an avalanche, to certain death and destruction: To overcome this danger, good-sized trees were attached by chains to the rear of the wagon, with the branches left on, to act like grappling-irons, and in this manner the descent was made in safety. But woe to the careless or the unlucky wight whose improvised "brake" became uncoupled. The best he could hope for, in that case, was that a fore-wheel *would* dash up a tree, even if an upset was the consequence.

It sometimes happened that the oxen struck their heads against a solid fir-trunk; in which case, their proprietor became suddenly *minus* that pair of oxen, and *plus* a great many fragments of wagon and contents. Notwithstanding which pioneer incidents, very good roads now exist over the mountains in various places.

There is no dry season on the summit of the Cascade Range; hence trees that belong to the coast region re-appear above the region of firs, such as the black spruce, which, fed by the sea-fogs that drift over from the sea and are caught in the mountain-tops, grow abundantly. Pines, larches, dwarf junipers, and occasional cedars also flourish at a height of over five thousand feet. Looking at the mountains from the valley of Western Oregon, we see no bare peaks until we come to the snow-line. The numerous snow-peaks seem to shoot up out of evergreen forests: the more so as all the snow-capped mountains rise from the eastern side of the range.

The ascent of the snow-peaks from the western side is necessarily attended with much difficulty, except in the case of Mount Hood. The road before referred to, as leading to the Dalles, passes around the base of this grand mountain. At Mountain Meadows, the highest point on the Dalles road, we seem to be just at the foot of it, where, disengaging itself from the company of lesser peaks, it springs up clear and free, a pyramid of rock and ice, thousands of feet higher than its neighbors—bold in outline, clear-cut, symmetrical, inexpressibly grand.

For a little distance above the meadows, the mountain appears belted with a dark girdle of trees. Above and beyond that, all is sharply defined in white and black; glistening snow-fields reaching up and up, scarred here and there with projecting needles and cliffs of basalt, or seamed for immense distances by rocky ridges and yawning chasms of blackness. So cold, hard, and immovable it looks, that it is difficult to attribute to its volcanic forces the upheaval of this vast basaltic and plutonic mass over which we are traveling.

But this frozen aspect is a deceitful one, as we are aware, since our own eyes have beheld the fiery column shooting up from the old crater, followed by great volumes of dense, black smoke. The grand old mountain is not often stirred in these centuries of peace ; but it holds within its bosom fires that have never gone out since the morning of creation.

The ascent of Mount Hood, owing to the road, is not difficult. Every summer parties go up it, and many memorials are deposited there of these visits. August, or the latter part of July, is the most favorable time to make the ascent, when the snow is neither too hard nor too soft. The earlier one can go with safety, the better ; as there is likely to be, late in the season, a good deal of smoke from burning forests, which obscures the view. In clear weather, the panorama which can be enjoyed from the summit of Mount Hood is worth a journey across the continent to behold.

Mount St. Helen, though in Washington Territory, is reckoned among the Oregon mountains, because it is visible not only from the Columbia River, but from the heart of the Wallamet Valley. Not so high as Mount Hood, it is remarkable for the symmetry of its rounded dome. It is not difficult of ascent, except on account of the intervening forests. It is approached by following the north fork of the Cathlapootle or Lewis River, which enters the Columbia opposite the mouth of the Lower Wallamet. As the melting of snow in the mountains swells this stream to a rapid torrent in the early part of summer, the undertaking

21

must be postponed to the last of the warm season. Even then it is a very rugged and dangerous trip, though it has been accomplished by a few old mountain men.

One of these related to us how, while endeavoring to reach a certain bald, black spot on the west side, known as "the bear," he lost his footing, and went, as he expressed it, "kiting" down the side of the mountain, expecting nothing else but to be dashed to pieces. Fortunately there are few crevasses on this mountain ; and, coming to softer snow, he was able to check his speed, and regain his footing. He found that "the bear" was a black rock, kept *bare* by hot springs, which burst out at this place.

Mount St. Helen has been frequently known, since the settlement of the country, to throw out steam and ashes ; scattering the latter over the country for a hundred miles, and obscuring the daylight (on one occasion) so that it was necessary to burn candles.

Fine gold is found in such quantities on the Cathlapootle River that many attempts have been made to prospect at the foot of this mountain. But these attempts have always been frustrated by the obstacles already mentioned. In time the mineral wealth of the Cascade Range will be developed ; not, however, until the population has been greatly augmented, and the necessary gradual clearing up of the country opens the way.

Mount Adams, almost directly east of Mount St. Helen, and visible from the Wallamet Valley, like the Oregon snow-peaks, rises from the eastern side of the range, and can be reached from the more open country on that side without any great exertion. It is one of the most imposing of the snow mountains in appearance, although not one of the highest. The best views of it are to be obtained from the hills near Dalles City.

From the Columbia and Wallamet rivers one gets just a glimpse of Mount Rainier, the grandest peak of the Cascade Range, being 14,444 feet in height. Seen at this distance, and obstructed by St. Helen, no proper idea of its magnificence can be obtained. It is only when the divide which separates the Cowlitz River from the Puget Sound country has been passed that its beautiful proportions can be estimated.

From the prairies south of the Sound, it seems to have a triple summit; this appearance being caused by the wearing away of the mountain about its craters, two in number. That it is an immense upheaval is evident from the breadth of its base, which is twenty-five or thirty miles. It has been ascended, with great labor in getting to its foot. Above the region of forest are beautiful green meadows, spangled with flowers of the most brilliant dyes, dotted over with small groves of balsam fir. In the depressions between these green ridges snow lies, even in August, making a charming contrast with their emerald brightness; and above them towers the broken, icy pinnacles of Rainier. From its summit may be seen the glaciers filling its gorges, crossed again and again by deep crevasses.

Mount Baker is another lofty snow-peak of Washington Territory, though so far north as to be seen only from the Sound, or the Straits of Juan de Fuca. More active as a volcano than the other peaks, it has suffered loss of height and change of form, consequent on the falling in of the walls of its crater, within the last five years. This mountain, too, has been ascended—an interesting account of which appeared in *Harper's Magazine* about two years ago.

About centrally situated, with regard to the Oregon division of the Cascade Range, is a group of snow-peaks called the Three Sisters, which may be ascended without difficulty from the eastern side. Indeed, in order to get a well-formed idea of the shape of the mountains it is necessary to see them from this side.

Starting from the Dalles, and keeping toward the south until we strike the Des Chutes River at the Warm Springs Reservation, we find ourselves directly abreast of Mount Jefferson, with a complete and beautiful view of it. There is no labor in traveling over the piney slopes of the mountains here. It is more like riding through interminable parks, with little or no under-growth, a dry soil, abundance of flowers, and occa-sional small game. Three or four days' easy travel, through a country abounding in natural wonders, brings us to the Three Sisters.

They stand in a triangular group, the base of the triangle being toward the west. Though perfectly dis-tinct peaks—the northernmost being highest—they are connected near their base by lesser intervening ele-vations. Accustomed as we have become to mount-ains, the Three Sisters force from us the profoundest expressions of admiration and delight. So lofty, so symmetrical, so beautifully grouped! Nor are there wanting adjuncts, which augment the interest of the scene. At the foot of the group stands a single needle of basalt several hundred feet in height, in its grim, black hardness looking like a sentinel guarding the Olympian heights above.

We prepare to ascend the north Sister. By reason of the greater general elevation of the country on the eastern side of the Cascade Range, and the more grad-ual slopes also, the toil of an ascent is greatly diminished. By keeping along a ridge we find it comparatively easy to clamber up. Two of our party, however, decide to attempt a more abrupt ascent.

As we course along our rocky ridge we watch the adventurers on the snow-field. After climbing over a sharp slope of broken rock, they come upon an incline of nearly eighty degrees—in fact, the snow-field appears concave to us—and commence crawling up it. By great exertions, and cutting steps in the snow with their hunting-knives, they reach the edge of the first crevasse, where we see them pause, holding on to the edge, and looking into it. They can proceed no farther. The crevasse is fifteen feet across, and hundreds deep. Could they throw themselves over, they must inevitably slide back into it, from the glassy surface above.

Starting cautiously to return, and holding back by striking their heels in the snow, making but slight impressions, first one, then the other, loses his hold, and down they go—swiftly, swiftly, ever more swiftly—darting like arrows from their bows, straight down the steep incline, toward the rocks below the snow-line. The younger and more active contrives to draw his hunting-knife from its scabbard, and, by striking it into the hard snow, to check his speed. What a grip he has! We laugh, while we are trembling with excitement, to see him swing quite round the knife-hilt, like a plummet at the end of a string swung in the fingers. He has arrested his descent in time to avoid the rocks.

Not so his clumsier companion, who comes down, luckily, heels foremost, among the rocky *debris* at the bottom. His bruises, though many, are not dangerous; and this little experience teaches our young

friends the needful prudence. They are content thence-
forth to take the "longest way round," which is the
surest way to the object of their desires. After two
or three hours of clambering, we reach the line of perpetual snow.

Just below it is a belt of cedars, with tops so flat
that we walk out on them a distance of twenty feet,
either side their trunks. Early in their struggle for
existence their tops have been broken off by the wind,
and the weight of many winters' snows has retarded
their upright growth, until the result of a century of
aspiration is a ludicrously short stump, and immensely
long and broad limbs. In this region we find a few
stunted "mountain mahogany" trees ; but we are quite
above the pines.

Above this, in the snow, or rather in the thin layer
of soil deposited in places among the rocks where the
sun's action prevents the snow from accumulating, are
several varieties of flowering plants with which we are
familiar ; the blossoms, however, are but the miniature
copies of their valley kindred. So fragile, of such
delicate hues are they, that a feeling of tenderness is
inspired by their lonely position on this bleak summit ;
and we ask ourselves : For whose eye has all this beauty
been spread, age after age, where human footsteps never
come? Let those who believe every thing terrestrial
"was made for man," search those places of earth
where only God is, and study their adornments.

The view from the peak of our mountain is one long
to be remembered. To the north of us stretches the
Cascade Range, with its wilderness of mountains, from
six to eight thousand feet in height, overtopped by
Mount Jefferson and Mount Hood. To the south, the
same wilderness of mountains is seen over the tops of

the other Sisters, with Diamond Peak, South Peak, Mount Pitt, and, far distant, one which we fancy may be Shasta.

To the east, spread away immense plains, with their river-courses marked as on a map, and bounded by the Blue Mountains. Just below is the Des Chutes, and on the other side of it, not far off, is the extinct crater of a volcano, its remaining walls being only two or three hundred feet high. All around it the country is covered with black cinders, ashes, and scoria. Turning toward the west, we behold the lovely Wallamet Valley, with its numerous small rivers, its hills and plains, and beyond it the blue wall of the Coast Mountains.

We resolve to return to the pine woods to camp, and with to-morrow's dawn to climb once more to the summit, to behold "morning on the mountains." The spectacle compensates for the extra toil. When we arrive, there is a veil of mist hanging between the valley and the mountain-top. We know that they in the valley see nothing of the summits; while we of the summits can discern nothing below this floating sea of vapor. How beautiful! It is as if out of a sea of golden-tinted mist are springing islands of dark-green —some of them crowned with glittering snow—and overhead a cloudless heaven. With every moment some new and beautiful, but almost imperceptible, change comes over the misty ocean in which are bathed those isles whose shores are abrupt mountain-sides; and, in turn, all tints of gold, rose, amber, violet, float before our enchanted eyes.

Not long the scene remains. An August sun quickly disperses the gossamer clouds, unveiling for us the scene of yesterday in its morning sharpness of outline, with high lights and deep shadows in the foreground, and with a soft, illusory glimmer in the deep distance. We hardly wait for the full blaze of day on the picture, preferring to remember it in this more striking aspect.

Along the crests of the mountains are frequent lakes, some of which occupy old burnt-out craters; others may have been formed by the damming up of springs by lava overflows; others by a change in the elevation of certain districts, leaving depressions to be filled by the melting of snows, or by mountain springs and streams. These lakes occur generally where signs of recent volcanic action in the neighborhood are most numerous, as in the vicinity of Mount Jefferson, the Three Sisters, and Diamond Peak.

Pumice, cinders, scoria, and volcanic glass, with other evidences of eruption comparatively recent, abound all along the eastern base of the Cascade Range, and extend some distance through the central portion of Eastern Oregon. The traveler and scientific man must ever be amply repaid for the labor of exploring the country east of the mountains, by the great and varied wonders which meet him at almost every step of his journey.

It does not prejudice a country, either, that it is of volcanic formation; for, wherever the soil has had time to form, it is sure to be of that warm and fertile nature that produces every thing in abundance, and quickly. Probably the eastern slopes of the Cascades will sometime be celebrated for their grapes and peaches, as now the foot-hills of the Sierras are. In both instances, the soil and climate are identical.

CHAPTER XXXI.

WE owe to Rev. Thomas Condon, of Dalles City, all our real information upon the geology of Oregon, as well as many notes upon its mineralogy. That which we were able to observe for ourselves only corroborated his views. According to Mr. Condon, the Rocky Mountains once formed the western breakwater of the continent, as the Coast Mountains now do. They were forced up by the subsidence of the ocean-bottom, and the consequent upfolding of the earth's crust. The upheaval occurred near the shore-line, but left a narrow strip of the old sea-bed east of the Rocky Range; enough to prove that the upheaval occurred in the Cretaceous period. A large body of salt water was thus isolated, which gradually, by natural drainage, became brackish only, and finally quite fresh. This change is also proven by the nature of the deposits.

After a long interval of quiet, another upheaval took place, occasioned, like the first, by a subsidence of the ocean-bed. At this second folding of the earth's crust, the Cascades and Blue Mountains were forced up, and once more a large body of sea-water was divided off from the ocean, to form great salt lakes, which gradually became fresh. The Blue Mountains formed an island, separating the northern portion of these waters from the southern, which were drained respectively by the Columbia and the Colorado rivers; but not until by deposits of various character the bottoms of these basins became sufficiently elevated.

In like manner, the later upheaval of the Coast Range caused to be inclosed between these mountains and the Cascade Range another immense body of water, which became fresh in time like the older lakes, and with the gradual elevation of the sedimentary deposits was finally drained off like them. That the dates of the formation of these lakes were widely separated is evident from the fossils of each, which indicate the geologic period to which they belonged; the deposits of the Wallamet Valley being the most recent.

In the meantime, vegetable and animal life flourished along the shores of these inland seas, or lakes. There are canyons in Eastern Oregon, fifteen hundred feet in depth, whose walls present a complete and undisturbed record of the geologic periods. First of all in this record is the old ocean-bed of the Cretaceous period, teeming with myriads of marine shells, perfectly preserved in form, though frequently containing, as a mold, a filling of chalcedony or calcareous spar, making specimens of the highest beauty.

Next above the salt-water deposits, come those of the earlier Tertiary periods. In this division, we find the leaf impressions of those grand trees that flourished during ages of tropical warmth and moisture: palms, yew-trees, immense ferns. In some places an oak-leaf, or an acorn-cup, has left its print in the rocks.

Contemporaneous with the palms and ferns were two species of rhinoceros, and three or four species of

Oreodon, an animal allied in some things to the camel, and in others to the tapir family. Another animal of a tapir-like appearance, but called by geologists *Lophi-odon*, also lived during this period, and left his bones in the muddy lake-margins to become part of earth's history. Also, a peccary of large size, and an animal bearing some resemblance to the horse, called the *Anchitherium*—found also in France, and in the *Mauvais Terres* of Nebraska.

Following this age, was one of volcanic action and the outpouring of immense quantities of ashes and lava. By the lava-streams issuing from the Blue Mountains new barriers were raised, dividing the northern portion of the great lake of Eastern Oregon more completely from the southern, which, by reason of superior drainage, was the first to become dry land. The lake on the northern side of the Blue Mountains remaining longest a lake, continued to receive the drift of its shores for a longer period, and consequently offers a more perfect record of the changes which took place through all the Tertiary periods. Several of the strata formed in this lake are of pure volcanic ashes, still rough as pumice-stone to the touch.

Thus, this Middle Tertiary period was closed in violence. Volcanic fire, earthquake-shocks, and molten lava destroyed and blotted out all forms of vegetable and animal life. The ages roll on, and once more living forms of plant and animal haunt the shores of these shallowing lakes. The oak, the yew, the willow, have left their prints in the sedimentary rocks; and the bones of new creations of animal life, such as the camel and the horse, accompany them. But these, too, in turn suffer extinction by violence; the whole country being covered more than thirty feet deep in volcanic ashes.   Indeed, deposits of volcanic ashes exist in Eastern Oregon which are one hundred feet in depth.

After a long night of geological darkness, during which there seems to have been a subsidence of earthquake and volcanic outflow, life once more appears upon this portion of the earth in the forms of elephant, ox, horse, and elk; accompanied by such vegetable forms as were suitable for their subsistence. But yet another period of death was to ensue before the frame-work of the present Oregon was perfected. And this time the desolation appears not to have come from fire, but from frost and flood.   How long it continued, or what mighty seas of ice moved over the face of the earth, marking the hardest rock with glacial abrasion, none can tell.   But to have so clearly written in the rocks of Oregon the geologic history of at least one continent, is most interesting to scientist and amateur alike.   So far as can be seen, the Columbia River Valley must become the most desirable field for the student of the earth's history; and also of research into the record of pre-historic man.   For here, somewhere hidden in these ancient pages of rock, must be the beginning of man's history preserved, like that of God's other creatures, in tablets of stone.

From the brief sketch of Oregon's geologic history which has been given, it will appear what the agency has been of those glistening white snow-peaks—Mt. Hood, St. Helen, Adams, Jefferson, and all the rest—in forming the Oregon and Washington of to-day. Time was when these mountains belched forth molten lava, and rained hot ashes over many miles of country on either side.   For some reason—perhaps the direction of the prevailing winds—the ashes were chiefly deposited on the eastern side of the range.   The volcanoes themselves, in general, stand on the eastern side of the summit of the range. A covering of lava, in the form of basaltic rock, conceals from sight the record we have referred to, except where by the action of water the pages of the book have been cut through from cover to cover—from ocean-bed to overlying basalt.

For a distance of sixty miles east of Dalles this last overflow may be traced, growing thinner and thinner, until it becomes a mere capping on the hills. Underneath it all is sedimentary, except the interruptions, several in number, of the older outflows of lava. It is owing to the large extent to which volcanic ash enters into the composition of the earth and soil of this portion of Oregon and Washington that both earth and water are so often strongly alkaline. It forms a soil inexhaustible in fertility, and particularly adapted to the growth of cereals; but owing to its elevation, and to the depth of the streams below the surface, together with a dry climate, is difficult of adaptation to the uses of the grain-raiser.

Apropos of the geological formation of Eastern Oregon, Mr. J. Wassen writes in the *Overland Monthly*, for February, 1869, the following:

"Coming from the north-east, the Blue Range of Oregon, the Cascade Range from the north, and the Sierra from the south, blend into or form a vast steppe or table-land of lava and sage-fields, interspersed with a score of lakes, in size varying from five to forty miles in length, and proportionate width. This high separating belt of land and water commences at the Owyhee River, and extends westward to the mountains, running at right angles to the ocean—a length of three hundred miles, and an average breadth of one hundred and fifty. There are three distinct chains of lakes in this district: The eastern, known as the Warner, inclusive of the Harney and Malheur. The second chain of lakes may be called the Goose Lake, including its northern links—Albert, Silver, and other smaller lakes. Goose Lake nestles in the extreme north end of the Sierra, and is the source of Pitt River, the main branch of the Sacramento. This fact has been disputed, owing, perhaps, to the outlet being underground in the drier seasons. The third and last, and larger of the several chains, is the Klamath, embracing Wright and Rhett lakes, farther south. The Warner lakes string along more like a river; and the rapid current, setting north at all times, is suggestive that this line of water is really the outcropping of a long, subterranean stream. The amount of water is apparently more than the natural drain of the country adjacent; and the outline of a great river channel is distinctly traceable to the lakes of Harney and Malheur. The latter, however, are strongly tinctured with the alkaline soil surrounding.

"The variety and great quantity of fish for which the streams feeding these lakes are noted; the myriads of water-fowl of every conceivable species that make these lakes their summer resort, and the countless numbers of deer, antelope, and the larger game, contribute principally to make the district of the lakes, what it surely is, the happy hunting-grounds of the expiring race. They are hardly to blame for the tenacity displayed in its defense; this broad pass in the mountains furnishing the wily savage with a hundred avenues of escape, to the right or left, with his plunder and his life. The shelving shores of the lakes furnished him warm winter shelter, and the great depressions natural trails free from snow· in the severest seasons. These trails are carefully flanked at favorable intervals with little bastions and semicircular breastworks of loose stones, mementoes of Indian skill and strategy. Aside from any known or prospective material resource, the district of the lakes, with its dense forests and weird deserts, picturesque mountains and delightful valleys, and silent waters inclosed by perpendicular walls of mysterious formation, must ever be a scene of enjoyment for the tourist and lover of all that is grand, beautiful, and peculiar in Nature."

Thus does the observing traveler confirm the views of the student of geological science. The southern half of Eastern Oregon retains yet some of the features of the *undrained* lake districts of Oregon and Washington.

That portion of Oregon and Washington which lies west of the Cascades is part of a great trough, extending from the Straits of Fuca to the Bay of San Francisco. It is not, like Eastern Oregon, elevated above the original sea-bed by immense deposits of volcanic matter; but its older rocks are buried from sight by deposits of the Tertiary and post-Tertiary periods.

There is a curious glimpse into the pre-historic record of man given by the fossils of the Wallamet Valley. For instance, the teeth and tusks of the elephant have been found in Linn, Polk, and Clackamas counties, at no great depth below the surface—as in three instances they were discovered by men engaged in digging mill-races, probably from eight to twelve feet in depth. Side by side with this fact, is the one that at a similar depth some rude stone carvings have been discovered, buried in the alluvial soil of the Lower Wallamet, about two miles above its junction with the Columbia, in Columbia County. Stranger still, there

has been discovered at a place just at the northern end
of Multnomah County, the remains of a camp-fire,
with the half-burnt brands lying in position, as if the
fire had but just gone out, and buried under *twenty-
seven feet* of alluvial deposit. Equally curious is the
fact that in the Nehalem Valley, eight miles back from
the coast, and twenty-five feet below the surface, in
a place where there is no suggestion even of a possible
land-slide, was lately discovered a large knife of pure
copper, with a stone handle. Here is a *souvenir* of the
stone and copper age! Shall we ever be able to col-
lect any facts concerning these ancient Oregonians?
The paleontologists have here a splendid field to
work in.

The work of the volcanoes is also very evident in
Western Oregon. The valley of the Lower Columbia,
in particular, reveals the immense overflows of lava in
its many forms of basaltic rock. In many places, it
occurs in solid masses of many feet in thickness; in
others, it has assumed the columnar form; and in many
more, it is broken into sharply angular fragments, mixed
with earth. The fracture in the latter case is foliated
—every fresh cleavage showing what appears like the
impression of palm-leaves. The most interesting form
of basalt occurs in some columns in the high river-
banks just below the town of St. Helen. These columns have been brought to view by the gradual process of denudation; and now project only a dozen feet
or so of their tops from the incline of the high bluffs.
They consist of uniform blocks, of about ten inches
in thickness, having six sides—laid one above another
so as to appear like a solid pillar. But their great peculiarity is, that each individual block has a similar-
sized chip off the lower side, on its north-west corner

or angle. With this exception the blocks are flat. Occasionally one gets thrown off, and so the columns never appear at any great height above the earth ; but their fragments strew the river-bank for a long distance.

This basaltic outflow evidently came from Mount St. Helen. On any of the sand-bars in the Lewis or the Cathlapootle River, which debouches into the Columbia on the opposite side, are to be found water-rolled fragments of pumice-stone in abundance; and there are seasons of high-water which bring down from Mt. St. Helen by some of its streams—the Cowlitz in particular—so much white volcanic ash as to render the water milky in its appearance. It is somewhat remarkable that while on the Oregon side the basalt covers every stratified rock or sedimentary deposit, on the Washington side the hills are immense deposits of coarse gravel or sand, and water-rolled stones.

About in the central portion of the Wallamet Valley are some gravel beds of no great thickness; while in Washington Territory, along the Columbia and in the Puget Sound region, the soil is gravelly to an extent which renders it almost unfit for cultivation. Did the facilities which the Sound offered for drainage prevent the deposit of soil-making matter during the period of its submergence?

There are evidences, in the elevated beaches of the Oregon and Washington coast, of great changes of water-level over that portion of these countries west of the Cascades. At Shoalwater Bay, for instance, where the action of the surf has undermined large portions of the bluff shore, breaking it off, there are, exposed to the eye of any observer, vertical sections of sedimentary deposit one hundred feet above the present sea-level. Mixed with this deposit, and sometimes occurring in beds, are vast numbers of sea-shells, of the kinds now common to our oceans. The presence of oyster, clam, and other shells, only found in shallow water; as also of trunks of trees, leaves, seeds, and cones—their forms preserved unbroken—proves these fossils to have been deposited quietly in water of no great depth, and to have remained undisturbed since. Granting this apparent fact, the waters in which they were deposited must have stood more than a hundred feet higher than the present level of the ocean; or enough higher than the highest of these deposits to have sufficiently covered them.

Mr. Condon's theory, which we have already adverted to, supposes what is now the Wallamet Valley to have been the basin of a large body of water, to which, in an article in the *Overland Monthly*, of November, 1871, he gives the name of the Wallamet Sound. The conclusion of that article has this interesting summing up:

"And now, with our amended theory in mind, as a measuring-rod, let us retrace our steps to the lower country—the Wallamet Sound of the olden time. Let the fall of the Columbia River, from this lake-shore east of the Cascade Mountains to the mouth of the Wallamet River, be stated at eighty feet. Our fossil remains on this lake-shore are 250 feet above the present level of its waters, making a total of 330 feet as the depth of those waters above the present surface at the mouth of the Wallamet River. How naturally one looks to the currents of such a vast body of water as the agency competent to the heaping up of that long, sandy ridge, one hundred feet high, through which the river has cut its way at Swan Island, north of Portland. But let us follow it still farther inland.   Over where Portland now stands, these waters were 325 feet deep; over Salem, 165 feet; over Albany, 115 feet; over Tualatin Plains, 145 feet; over Lafayette, 170 feet.   A narrow strait, over the present valley of the Tualatin River, ten or twelve miles in length, opened westward upon a broad, beautiful bay, extending over the present sites of Hillsboro and Forest Grove, to Gale's Peak, among the foot-hills of the Coast Range.   The subsoil of the fine farms of that rich agricultural region, is itself the muddy sediment of that bay.   Farther south, over the central portion of the present valley, and lying obliquely across the widest part of that Wallamet Sound, there arose above those waters an elevated island.   It extended from a point south of Lafayette to one near Salem, and must have formed a fine central object in the scene.   Three or four volcanic islands extended, in an irregular semicircle, where Linn County now is; and the islands of those waters are the Buttes of to-day—Knox's, Peterson's, and Ward's.   One standing on the summit of either of these Buttes, with the suggestions of these pages before him, could so easily and vividly imagine those waters recalled, as to almost persuade himself he heard the murmuring of their ripples at his feet—so sea-like, the extended plain around him—so shore-like, that line of hills, from Mary's Peak, on the west, to Spencer's Butte, on the south, and only lost, on the east, among the intricate windings of extended slopes among the foot-hills of the Cascades.   How natural would seem to him this restoration of one of geology's yesterdays!

"The shores of that fine old Wallamet Sound teemed with the life of the period.   It is marvelous, that so few excavations in the Wallamet Valley have failed to uncover some of these relics of the past. Bones, teeth, and tusks, proving a wide range of animal life, are often found in ditches, mill-races, crumbling cliffs, and other exposures of the sediments of those waters, and often within a few feet of the surface. Did man, too, live there then? We need not point out the evidences of increasing interest the world feels in facts that tend to solve the doubts that cluster around this natural inquiry. A few more mill-races dug, a few more excavations of winter floods — more careful search where mountain streams wash their trophies to their burial under still waters—and this question may be set at rest, as regards that Wallamet Sound. Oregon does not answer it yet."

Oregon has no State Geologist; and so far has been subject to the investigations, chiefly, of one man, Mr. Condon, with the exception of such slight observations as have been made from time to time by the Government explorers for the Pacific Railroad. That it is a field well worthy of scientific research there can be no doubt, nor that it is one which will richly reward the necessary outlay of money.

CHAPTER XXXII.

THE valuable minerals of Oregon are : first, the precious metals, gold and silver; and second, copper, lead, iron, coal, marble, and salt. There are also various earths and stones useful for manufacturing purposes, and doubtless minerals of greater value concealed in the almost wholly unexplored mountain ranges, which the further development of the country will bring into notice.

Concerning the formation of the metals, more especially of gold, there are many theories. The age of the rocks associated with gold must serve as an indication of some value in pointing out its origin; the most probable theory of which seems to be, that, at a period when great changes were going on in the shape of the earth, the upheaval of mountains and overflow of volcanoes, certain vapors contained in the earth being forced by heat and pressure into the fissures of rock already hardened, or even into the substance of rock not yet solidified, became precipitated in the form of gold upon the walls of the cavities which shut them in. Much of this gold was subsequently set free by the action of the water, and is found mixed with sand and gravel, or earthy matter, in old river-beds or valleys between high mountains. Much of it still remains in its original position, and has to be got out of the rock by blasting and crushing.

The gold-fields of Oregon lie along the bases of, or in close neighborhood to, its mountain ranges; and there is no mountain chain which has not somewhere along it a gold-field, more or less productive. As to the mountains themselves, in Western Oregon, their rugged nature and impenetrable covering of timber have prevented their being prospected. It is only in the placer diggings of the southern counties, and the beach diggings of the coast counties, that mining for gold has been carried on to any extent.

After the rush of '49 to the gold-bars of the California rivers had made miners and experts of a hitherto purely agricultural population in Oregon, they began to find indications on their own soil of the existence of the precious metal. Traveling overland to and from California gave them opportunities of observing the nature of the country, and it was not long before the gold-hunters stopped north of the California line. As early as 1852 good placer diggings began to be discovered, and for a number of years were worked with profit. They still yield moderately, but are chiefly abandoned to the Chinese miners, who content themselves with smaller profits than our own people.

Jackson County is divided into several mining districts, the gold being placer and coarse gold. Formerly some nuggets were found not far from Jacksonville, worth from $10 to $40, $100, and even $900; but no such discoveries have occurred of late.

The annual production of gold in Jackson County is a little over $200,000. About five hundred American miners and six hundred Chinese miners employ themselves in washing out gold dust. It will be seen, by averaging the amount produced among the number producing it, that it can not be, on the average, a paying business.

Of the quartz ledges, some of which are of undoubted richness, very few have been worked at all, and those which have, only very imperfectly. In one of these, the Gold Hill vein, a few miles from Jacksonville, $400,000 was taken out of a "pocket," after which the lead was lost, and the mine abandoned. From this pocket was taken a kind of gold peculiar to the deposit of the Cascade Range, called "thread gold." It is found in pockets, or basins, or chimneys of rotten quartz, occurring in veins of pure white quartz; and is really a mass of pure gold-threads, often in skeins, that, when examined under a glass, seem to be twisted—often arranged so as to resemble the bullion used for officers' epaulettes. It is very much matted together, holding in its tangled grasp small pieces of yellowish quartz. The same kind of gold is found in the Wallamet Valley, on the Santiam River; and in both instances is associated with free gold, embedded in hard, snow-white quartz. The specimens taken from these mines were very beautiful and extremely curious, and ought not to have been subjected to the crushing process; being worth more as specimens than as gold.

Salt, coal, and iron exist plentifully in Jackson County. Quicksilver is reported to have been discovered, but, so far, has never been worked.

Josephine and Curry counties furnish gold in about the same proportion to the amount of labor expended, that Jackson does. Its quartz leads have never been opened to any extent. One of the most promising mineral productions of these counties is copper, which, if not too pure to work to advantage, will yet make this portion of Oregon famous. Curry County embraces some of the most valuable beach diggings on the coast.

Coos County has also its gold, silver, and copper-mines of undisputed richness. But it owes most of its present celebrity as a mineral county to its coal, which, for several years, has sold readily in the San Francisco market; and the supply is apparently inexhaustible.

Douglas County has a gold-field situated on the Middle Fork of the Umpqua, and extending along the several creeks which head in the Cascade Mountains and their lateral spurs. Considerable gold has been taken out of the Middle Fork, Myrtle Creek, Cow Creek, and Coffee Creek diggings; and new ones are from time to time discovered. Silver is also known to exist in this county, though it never has been mined. Marble and salt are among its mineral productions; but its people being almost entirely an agricultural and pastoral community, little attention is given to any thing except farming and grazing. About where the north line of Douglas County intersects the Cascade Range a gold-mine has recently been opened, which promises to turn out very rich. Already a large amount of the precious metal has been taken out, and the indications continue to be good. This mine is called the "Bohemia."

There are no counties in the Wallamet Valley known as mineral districts. That gold and silver exist in the western slope of the Cascade Mountains is a well-known fact. So far it has been mined only on the Santiam River, where the famous pocket before mentioned was emptied of its contents. Many other lodes were located, but nothing has subsequently been done toward developing them. Other discoveries have been made nearer the Columbia River, in Oregon, and similar ones on the northern side of the Columbia, in Washington. Lead has recently been discovered in Linn County, near the Santiam gold-mines; and it must be regarded as inevitable that the base of the Cascade Range shall furnish in the future very considerable mineral interests.

The coal and iron of the Wallamet Valley, so far as yet discovered, is found at its northern end, either upon or within a few miles of the Columbia River. Limestone, which is very rare in Oregon, is found in Clackamas County, not many miles from Oregon City. Salt is found in Multnomah and Columbia counties, as also iron and coal. Black marble has been discovered in the mountains on the Washington side near the Lewis River, but has never been quarried. Some very good building-stone is also found in this locality.

Coal crops out frequently on both sides of the Columbia River, from the mouth of the Lower Wallamet to the sea. In the Valley of the Cowlitz there is an extensive deposit, of a good quality for fuel. Its steaming or gas-making qualities have never been tested. In appearance it resembles the Scotch cannel coal, burning freely when lighted at a candle, or in the open air. It has, notwithstanding, a woody structure, which places it among the lignites; and checks badly on exposure to the air. It has not, however, been worked sufficiently to afford a determinate judgment upon its commercial value.

Near the mouth of the Columbia, on the Washington side, at Knappton, is a cement factory. The scarcity of limestone on the north-west coast, and the cost of importing lime and cement from California, caused an enterprising firm of Portland to attempt the experiment of making the latter article from bowlders found in this locality, containing a considerable proportion of the necessary ingredients. The enterprise resulted in the production of a fair article so long as the supply of bowlders lasted ; but since the failure of this material the works have to depend upon a quarry of similar rock in the hills adjoining, and it is not yet determined whether or not the new cement will equal that produced from the bowlders. The capacity of the works is thirty-five barrels daily. Mr. Knapp, the energetic proprietor, has erected a most complete establishment, and intends to carry on his experiments to an anticipated success. Quite recently a silver lode of great richness is said to have been discovered in the vicinity of Astoria ; but the working of silver being so expensive, these rumors excite but little attention from that class of people who would have means to develop a mine of this kind ; yet the discovery may lead to the future development of mineral wealth in this vicinity.

There are in Oregon and Washington several factories for the manufacture of common pottery ; one at Vancouver, on the Columbia, and another at Buena Vista, on the Wallamet. There exist, in favorable localities, the best materials for the manufacture of fine earthenware ; clays of great smoothness and fineness ; and beds of volcanic substances, which, when fused, would evidently make a beautiful enamel.

The manufacture of salt was attempted, in 1867, in the northern end of Multnomah County, about half a mile from the Lower Wallamet River. The experiment proved highly satisfactory, so far as the quality of the salt produced was concerned ; but the capital required to make it a paying business has prevented its success in a financial point of view. Other salt-works, on a small scale, have been operated in Polk and Douglas counties. The salt made in Oregon is of a remarkable purity; so much so that a specimen of it was taken to the Paris Exposition by Prof. Wm. P. Blake, of the California Commission. The chemists of San Francisco pronounce it pure enough for the uses of the laboratory without being clarified, which no other salt in the market is.

Notwithstanding the amount and excellence of the iron ores of Oregon, they have never yet been made so profitable as they should be. The only works for the manufacture of iron are those located at Oswego, on the Wallamet River, six miles above Portland. The quality of the iron there made is said to be equal to the best Swedish; but the cost of its manufacture, arising from the high price of labor, and also somewhat from some ineligibility in the situation of the works, has prevented their entire success. A better situation for iron-works is on a bay of the Columbia, extending back of the town of St. Helen, in Columbia County, where extensive beds of ore exist in connection with coal, wood, fine water-power, and navigable water.

This is a brief account of the present mineral productions of Western Oregon, sometime, no doubt, to become famous for its manufactures, supplied by its home resources.

The eastern slopes of the Cascade Mountains do not, like the western, furnish gold-fields. Whatever treasure the laboratories of the far-distant past deposited in their depths, volcanic overflow subsequently concealed from human research, burying it under many successive layers of eternal basalt. The gold-field of

Eastern Oregon is to be found along the slopes and among the ridges of the Blue Mountains, where the marine fossils are not covered over by trap-rock. According to Mr. Condon, the older marine rocks, containing fossils of the *Rynconella, Cyrtoceras*, and other marine shells, may be considered indicative of the vicinage of gold-bearing rock.

The counties of Eastern Oregon, known as gold-producing, are Union, Grant, and Baker. The mineral districts are located on Powder River and Eagle Creek, in Union County; on the head-waters of the John Day, in Grant County; and on Burnt River and the head-waters of Powder River, in Baker County. Quite recently mineral discoveries have been made in the mountains about Goose Lake, in the extreme southern portion of Grant County, but have not yet been sufficiently worked to test their value.

Most of the gold produced in these counties has been taken from placer-mines, which generally have yielded well. Many quartz lodes have also been located, and a few containing free gold have been worked with good results. Quartz-mining has not, however, been carried on to any great extent in Eastern Oregon, the capital required to get out the ore and erect mills being wanting. It remains for wealthy companies in the future to undertake this order of mining.

Silver lodes, some of great richness, have been discovered in Baker County, a number of which are being worked, but not to any great extent. One is said to have yielded at the rate of over $7,000 per ton, by smelting on a common blacksmith's forge. What its working yield has been, we have not yet been able to learn.

Coal. iron, lead, and copper have been discovered in

the mineral districts of Eastern Oregon; without, however, exciting much interest, owing to the precedence given to the precious metals, as well as to difficult transportation, distance from markets, and other hinderances common to newly settled territories.

There are no very correct means of estimating the gold product of Eastern Oregon and Washington. The gold-fields of the north-eastern portion of the Territory have contributed a certain share to the general amount of bullion received by the Express-office and banks, which ship the gold to San Francisco; but it can not be separated from that of Oregon, Idaho, British Columbia, or Montana. It is only the gold of Northern Idaho that goes to the California Mint by way of Portland. All the gold of the southern portion of that Territory, and perhaps a part of that produced in Baker County, Oregon, goes to San Francisco by Wells, Fargo & Co.'s stages, overland.

There has been a regular decrease in the shipments of bullion by way of Portland since 1864, when the mining excitement in Idaho and Eastern Oregon was at its height. The shipments of Wells, Fargo & Co. from Portland have been as follows: 1864, $6,200,000; 1865, $5,800,000; 1866, $5,400,000; 1867, $4,000,000; 1868, $3,037,000; 1869, $2,559,000; 1870, $1,547,000. The shipments of Ladd & Tilton, bankers, of Portland, for 1869, were $419,657. There is always a considerable amount of gold dust conveyed by private hands at the close of the mining season, which can not be correctly estimated. Add to this the gold produced in Southern Oregon, which is about $400,000, and the sum total of all the bullion produced in Oregon and Washington will amount, for 1870, to about $2,000,000; whereas, it was probably $3,000,000 for the preceding year. This decrease is owing partly to the exhaustion of old diggings, and partly to the opening of other routes of travel, by which the gold dust is scattered in many directions, instead of flowing through one channel only, as in 1864.

There is no longer any excitement about mining in any part of Oregon, or the adjacent Territories. It has assumed the aspect of a steady industry, and as such will long continue to contribute to the wealth of the State, in connection with agriculture, manufactures, and every form of productive labor.

CHAPTER XXXIII.

ABOUT FARMING, AND OTHER BUSINESS.

The soils of Oregon and Washington have been already frequently mentioned as being a rich, sandy loam in the central valleys; a still richer alluvial, loamy soil in the small valleys of the mountain and coast regions, and a greater proportion of clay on the hills; while the soil of the great rolling prairies is fine and mellow, with considerable alkali in it. We have no intention of expatiating further upon their respective merits; but have thrown together clippings from the papers published in various parts of the country, from which the reader may be able to form an estimate for himself of their productiveness in general:

"Mr. Jake Greazier, of North Yamhill, sowed twenty-one acres of land in wheat, the yield of which was nine hundred and sixty bushels; an average of forty-five and two-thirds bushels to the acre. The yield of his oats was sixty bushels to the acre. We do not claim for all of Oregon as large an average yield as the above; but we do claim this, that when the land is properly farmed the yield will be near the amount above named. To be properly farmed, all wheat should be sown in the fall. In California they are adopting nearly altogether summer fallowing, and find that they are more than repaid in so doing."—*Portland Oregonian.*

"We 'call' you, *Mr. Oregonian.* In this county, Mr. Christian Mayer has harvested fifty acres of wheat, which yields sixty-four bushels to the acre; Stephen Brinkerhoff, thirty acres, averaging sixty bushels, and Orley Hull, eighty acres, with an average of forty-six bushels. We, of course, can not vouch for these statements, as we did not see the grain measured, but we take the word of the gentlemen and their neighbors. They are all well-known farmers here, and any one who thinks the figures too large can find out by asking for themselves. We do not wish to boast, but we claim the 'belt' for Walla Walla as a wheat-growing country."

"DOUGLAS COUNTY.—The *Plaindealer* says that S. C. Moore, who resides on the South Umpqua, five miles south of Roseburg, this year cut a field of wheat, containing eighteen acres, which yielded forty-five bushels to the acre. The land and grain were both accurately measured. The waste caused by some of the wheat being down, was estimated at five bushels to the acre. The *Plaindealer* thinks this a good crop, taking into consideration the fact that the season has not been a very favorable one."

" STILL BETTER.—Mr. Bleachleg, who lives about ten miles below Eugene City, says the *Journal*, has harvested this season an average of over fifty bushels of wheat to the acre, and from four acres a yield of over sixty bushels to the acre."

"The Albany *Register* states that a field of one hundred and fifty acres in Linn County yielded 8,250 bushels of the finest quality of wheat, the average being fifty-five bushels per acre."

"A field of wheat of sixty acres, belonging to Mr. Blackley, of Lane County, averages fifty bushels to the acre. Four acres yielded sixty bushels to the acre."

"The eastern country is boasting of mammoth squashes. The Baker City *Democrat* speaks of one weighing seventy-three pounds. The Winnemucca *Register* has seen one weighing seventy-five pounds. And now comes the Owyhee *Avalanche*, 'raising' its contemporaries by declaring that one is on exhibition in its town which weighs one hundred and six pounds. As the *Avalanche* had the last say it would have been its own fault if it had not told the story of the largest pumpkin."

"OREGON CHERRIES.—From the San Francisco *Alta*, of July 30th, we copy this:

"'A branch from a cherry-tree was shown to us yesterday which was certainly a little ahead of the average. It measured five feet in length, weighed seven pounds, and had three hundred and fifty-four cherries upon it. The variety is known as the 'Royal Anne.' It was from Seth Luelling's nursery on the Wallamet, near Portland, Oregon. It will remain for a short time on exhibition at Steele's drug store, Montgomery Street.'

"The branch of cherries from Mr. Luelling's nursery, now on

exhibition at Ferry, Russell & Woodward's, corner of Front and Alder streets, is only twenty-two inches long, and weighs over five pounds. There are too many cherries on it to count, but a *thousand* would be a safe guess, we think, and they are all huge 'Royal Annes.' If that branch astonishes the San Franciscans, they ought to see the one we speak of."—*Portland Bulletin.*

"I have seen large fields of wheat average fifty-six bushels to the acre, and weigh sixty-two pounds to the bushel; and have seen fields which yielded forty to fifty bushels per acre, from a 'volunteer' crop; that is, produced the second year from grain shattered out during harvest, sprouting during the fall, and growing without even harrowing. We generally raise the variety known as 'Club,' and sow it in the fall or spring. We produce about forty bushels of corn to the acre, of the large Yellow Dent variety, and it ripens nicely by the first of September. The potato is perfectly at home here, growing large, fine, and mealy. I let a neighbor have nine pounds of the early Goodrich variety, last spring, from which he raised 1,575 pounds. Sweet potatoes yield finely, but they are not so sweet as farther south. Turnips, beets, cabbages, tomatoes, peas, beans, onions, are all raised with ease and in great abundance. Although the country has been settled but a few years, there are already a number of fine-bearing orchards. I commenced here six years ago last spring, on ground that had never been fenced or plowed. After thoroughly plowing up about five acres of ground, I planted it in orchard with small yearling trees. This year I had one thousand bushels of the finest peaches that I ever saw grown—fully equal to the best Delaware and New Jersey peaches—besides large quantities of apples, pears, plums, cherries, apricots, grapes, and every variety of small fruits. Fruits of all kinds are perfect in every respect in this climate, particularly plums, the curculio having never been seen. I have one hundred bearing plum-trees. One Imperial Gage, two years ago, produced four hundred pounds of delicious, rich fruit, which brought eight cents per pound in gold; last year it had about the same amount of fruit, which sold for twelve and a half cents per pound, gold; many other trees did nearly as well. There are a large number of orchards just coming into bearing in this country, which will, of course, bring down the price of fruit."—*Philip Ritz, of Walla Walla.*

"The Department of Agriculture, in the report of 1870, puts the average wheat yield of Oregon at twenty bushels per acre,

23

which was, according to that report, one bushel higher than the yield of any other State. Minnesota came next, with a yield of nineteen bushels. The fact is, that good tillage in the Wallamet Valley will obtain an average yield of thirty bushels to the acre, one year with another. The records of our agricultural societies show that premiums have often been given on wheat fields yielding forty, fifty, and even sixty, bushels per acre; and that from sixty-three to sixty-seven pounds per bushel is not uncommon. In Marion County the average in an entire neighborhood, one year, embracing a dozen or fifteen farms, was ascertained to be as high as thirty-four and a fourth bushels per acre. The wheat of the Wallamet Valley is of a superior quality. It contains more gluten than wheat raised anywhere on the Pacific Coast; and on that account, the flour made from it commands in San Francisco, where its quality has become known, a higher price than any other, among bakers and large hotel-keepers, for it is more profitable; it makes a greater weight of bread to a given quantity of flour.

"Oats is the principal crop raised for feed in the Wallamet Valley. It is always a sure crop, yielding from fifty to one hundred bushels per acre. It weighs usually from thirty-six to forty-three pounds per bushel, and commands from ten to fifteen cents per hundred pounds more in the San Francisco market than California oats. Corn and barley are cultivated in the Wallamet Valley to some extent, and good crops of both have been raised, yet they are not well adapted to the climate. In some particularly warm localities corn is raised every year for fattening hogs. The bulk of the pork, however, is fattened on wheat. It is a cheaper feed, and, with experience in curing, has been found to make equally good meat. Rye and buckwheat are good crops; the former yielding from twenty to thirty bushels per acre, and the latter from forty to fifty bushels. These crops do best on the hilly lands of the valley. They are always sure. The rye and buckwheat flour of the Wallamet Valley is superior to that of any other part of the Pacific Coast.

"Timothy, clover, blue-grass, and several other varieties of grass, are cultivated throughout the valley for hay and feed. Timothy is the principal dependence for a hay crop. On the rich bottoms and swales it yields from two to four tons per acre, and is always a sure crop. The native grasses of the valley furnish excellent pasturage, summer and winter, for stock of all kinds.

They are not equal to the grasses of Eastern Oregon for making beef, but they can not be excelled for dairy purposes, especially on the bottom-lands of the Wallamet and Columbia, in the northern part of the valley. These overflow in June, every year, from the melting snows in the mountains, that swell the Columbia beyond its banks. After the overflow subsides the grass comes forward quickly, furnishes a crop of hay, and then pasturage the ensuing fall and winter. No grass in the world, wild or tame, is better adapted to making butter and cheese. The time was, only a few years ago, when Oregon did not make good butter enough for home consumption. Even now a vast amount of butter is made here and shipped out of the State, that is a disgrace to the name. It is because very many people do not know how to make a good article, and they are too careless and indifferent to learn, or to appreciate a good article of butter when they see it. Within the past few years a few men have taken hold of the butter business near Portland, and are making a splendid article of butter, demonstrating that the wild grass of this section has no superior for that purpose.

"The fruits best adapted to the soil and climate of the Wallamet, are, apples, pears, plums, cherries, quinces, currants, and all the different kinds of small fruits, strawberries, blackberries, etc. Among these the apple is the staple. Like wheat, it is a sure thing anywhere in the valley, where the land is not positively swampy. There is generally a fair market for apples in San Francisco, as those of California production are of inferior quality; and, as the apple is a fruit that bears transportation, it will continue to be cultivated for export to a considerable extent. The hilly portions of the valley seem to be better adapted to fruit-growing than any other. The extensive tracts of timbered lands in the northern part of the valley are especially good for that purpose. These are convenient to the Portland market, and, also, to shipping facilities; hence it is probable that the fruit business in future will receive more attention here than elsewhere in the valley.

"The peach and kindred fruits do not succeed well, unless, it may be, in sheltered localities in the upper part of the valley. The climate is not adapted to their growth. Like corn, such fruits require hot weather and hot nights to bring them to maturity, instead of which we have in the Wallamet Valley the cool sea-air and cool nights all summer. There is too much moisture in the atmosphere; in short, the peach is away from its natural home on the soil of the Wallamet Valley. About as much might be said of the grape, although some varieties seem to do tolerably well. The vine is healthy, but the berry is subject to mildew. It will not mature every year. It has not the flavor that belongs to the grape in its native home. Some persons, however, have had good success with grapes. Very much, no doubt, depends on the locality. Cherries and plums are produced in great profusion and variety, particularly in the northern part of the valley, where the Portland market affords ready sale for them. In and around Portland, in the gardens and small farms, the culture of strawberries, blackberries, currants, and that class of fruits has become quite a business—one, too, that is increasing every year. The climate is adapted to their growth. They mature well, and yield very heavily. They are remarkably fine-flavored, very large, and otherwise of good quality.

"The nurseries all over the valley, as well as elsewhere in Oregon, are well stocked with every variety of fruit. Great pains are taken by those in that business to keep up with the demand for their products, to introduce new varieties, and improve the fruit of the country. Fruit-trees come into bearing much earlier than anywhere in the States of the Atlantic coast. Usually the third year from transplanting the tree begins to bear; at about six years old it is in full bearing.

"The Wallamet Valley may be relied on by the farmer as a safe place in which to pursue his vocation. Its products are the leading necessaries of life. The land is good; the climate mild and healthful; markets good, of easy access, and always reliable. The question has been frequently asked 'if the soil of the Wallamet does not wear out?' It has never yet worn out, and some of it, too, has been in cultivation continuously for twenty or twenty-five years. The crop is apparently as good now as that of fifteen or eighteen years ago. The annual wheat yield is, approximately, about three million bushels; of all other grains, about 1,500,000 bushels. A very large amount of live stock, dairy products, wool, and bacon are marketed annually, the cash value of which can only be approximately estimated, say about $2,000,000."—*Democratic Era.*

"In our last issue, we placed the stock of flour on hand July 1st, 1871, at 78,500 bbls., chiefly Oregon brands. We have since seen a statement that the flour stock in Portland, Oregon, at same date, was 15,000 bbls. During the past harvest year we received from Oregon 179,536 bbls.; from the interior of this State, 123,513 bbls.—the former 53,000 bbls. more than the year previous, and the latter 61,000 bbls. less. Now, then, how much flour was manufactured in this city the past twelve months? Probably not less than 200,000 bbls., or considerably less than for the previous harvest year. Our millers do not like to see the Oregonians successfully competing for their legitimate trade, and measures to check the same are being taken, which will doubtless effect a serious change in this regard; improved patent machinery being brought into play, thereby greatly reducing the cost of manufacturing. Oregon has already opened a direct export wheat trade with the United Kingdom, and more or less flour has also been exported to foreign marts, and it is to be presumed that some increase in this direct trade abroad will naturally result. Vessels bringing cargoes of railroad iron, salt, etc., to Oregon, will naturally carry off cargoes of both flour and wheat in exchange, though it is believed that upon the completion of the Northern Railroad to Oregon, much of the produce of that State will be diverted inland, while more or less will seek this market by rail. Heretofore the Oregon farmers have been so isolated from available markets that the millers of that State have had large advantages during the long winter months, buying up wheat at very low rates, grinding it at their leisure, and shipping it to the most available markets, and at such a low cost as to defy all competition. Changes of some moment in this regard are imminent, and that at no distant period."—*San Francisco Commercial Herald.*

"WHERE THE SHOE PINCHES. — A San Francisco exchange is responsible for the following, which is so eminently characteristic of California ideas that we publish it: 'Our city millers will certainly be glad to have Oregonians ship their flour direct to foreign marts, rather than send it here, to glut our market and keep up a sharp competition for the trade of this port. Our millers and shippers would much prefer to receive the Oregon wheat than the flour, for in the former case it can be utilized to a good purpose by mixing with the California product.'"—*Oregonian.*

"SHEEP BREEDING.—The high prices obtained for wool by our growers will doubtless lend an additional impetus to the raising of sheep for wool. The range in the Wallamet Valley and also in Southern Oregon is gradually closing up, forcing wool-growers to take their flocks to Eastern Oregon and Washington Territory, thus bringing the sheep into a drier climate, which will eventually, when shown to be more profitable, cause the breeding to a considerable extent of a finer breed of sheep. Our wool-growers have not bred up to the standard attained by California wool-growers; and, though our wool brings high prices, owing to its being of a year's growth and answering for combing-wool, still, we think, after being forced to go east of the Cascades they will find it profitable to give more attention to breeding for fine wool. In a climate such as is found east of the Cascades, experience has shown that fine-wool merino sheep do better than any other kind. Our factories have had difficulty in obtaining fine wool enough in Oregon for their use, and consequently have been forced to obtain supplies from San Francisco. The demand for a few years past for long wool induced many wool-growers in the older States and also in California to breed for long wool, but that demand is likely to be met by wool-growers East, who are given the preference owing to the wool being of a finer texture; but for fine wool there always exists a demand at round figures, and, let the supply be ever so large, still it will find a ready market."—*Democratic Era.*

"STATE AGRICULTURAL SOCIETY.—The exhibition served to show that, whatever may be the capacity of the State, the dairy interest is not fully developed. The cattle-yards exhibit the same facts. There were a few very nice cattle—Durhams and Devons, and grades from these, but not one animal had been bred for the milking quality. In fact, I am led to believe that such are not in the State.

"The Oregon farmer prides himself on his horses, and certainly many not without good cause. The improvement of the stock is made the excuse for the race-track for the 'trial of speed,' as it is termed. It is certainly a misnomer to call these races a trial of speed. They would more properly be known as a trial of skill in horse-jockeying, of attempts to deceive the bystanders—in a word, a perfect gambling-shop. While telling the truth in this matter, let me exonerate at least a large majority of the gentlemen managers of the society and fair. Their aim has been, and is, higher than merely to draw a crowd and give them an opportunity of betting. In the matter of this race-track they have an elephant on their hands, that is yearly degenerating the character of the fair, which in part will account for the meagre display in many departments, and which will eventually cause many of the more thoughtful people to withdraw their support altogether.

"The sheep husbandry was well represented by five or six different breeds, making a display well worth seeing. An enterprising California breeder introduced some Cotswold sheep, the first, I believe it is said, in the State; also some Cashmere goats that attracted universal attention. The South Down seemed, however, to be the favorite of many. In this department, there seemed to be great interest manifested, showing that the recent high prices of wool will again attract the attention of the farmer."—*Oregonian.*

"The census reports show the following agricultural products for Oregon for the year 1869:

	Bushels.	Value.
Wheat	1,750,000	$1,500,000
Rye	5,200	5,200
Oats	500,000	270,000
Corn	200,000	200,000
Barley	200,000	200,000
Potatoes	500,000	300,000
Hay (tons)	75,000	637,500

"The returns show that there were in the State, in 1869, 47,800 horses, 1,500 mules and asses, 79,312 milch cows, 101,500 young sheep, 112,700 swine, and 140,500 young cattle. Total value of domestic animals, $7,936,255. Cheese was produced the same year to the amount of 105,779 pounds, and butter, 1,000,159 pounds. The production of the State has been prodigiously stimulated since 1869 by the building of railroads and accessions to the population."—*San Francisco Bulletin.*

From these various quotations from home papers all may be gathered that is necessary to show the rapidly growing agricultural interests of Oregon; as well as to betray what branches of farming are neglected, to the injury of the State. When the *Oregonian* asserts that no cows are bred for their milking qualities, it confesses a great error on the part of Oregon farmers, and points out to the ambitious immigrant a new source of profit. With butter seventy-five cents a pound in San

Francisco, and from twenty-five to forty-five cents in Portland, there is money in the dairy business.

Fowls, which generally do not do well in California, are healthy and prolific in Oregon. Eggs always command a high price in San Francisco, and are by no means so cheap as they should be in Portland, simply because farmers neglect to raise hens.

The cultivation of flax is an industry only of late resorted to in Oregon, though the culture of this plant is highly profitable; and it is, besides, indigenous to the soil in many parts of this State and Washington Territory. We notice, however, the receipt of twenty thousand bushels of flax-seed at the oil-mills in Albany, this year, and the Pioneer Oil-mill of Salem must have received as much more.

Owing to the drought in California, wheat has sold in Oregon at from $1.00 to $1.45 per bushel the present year (1871); oats, at eighty cents per bushel; and hay, twenty dollars per ton. Wool brings from thirty-five to thirty-six cents per pound. The spring clip amounted to over two million pounds.

The Oregon City Woolen Mills, this year, shipped to Boston fifty thousand pounds of their surplus wool, and expected to ship one hundred thousand pounds more.

The San Francisco *Bulletin*, in a review of things seen at the Mechanics' Institute Fair in that city, gives a very favorable notice of Oregon City woolen goods, and adds: "The production of this company consists chiefly of tweeds, flannels, cassimeres, blankets, and yarn. They manufacture mainly for the Oregon market, which gives little demand for the finer styles of woolen goods. In cassimeres, they claim superiority to any institution on the coast. Their stock of cassimeres is large, and of varied assortment. Many of the best pieces of goods seen in the windows of our aristocratic tailor-shops, are from this factory. Of tweeds, for men's and boys' wear, their supply is good. They make all varieties of blankets, from the coarse gray to the finest lambs'-wool, at forty dollars per pair. Their flannels are substantial, and of various grades of fineness. They exhibit forty-five pieces of cassimere, which those who love to examine good fabrics will be pleased to look at."

Farm hands can not be hired for less than $25 to $30 per month and board. Chinamen are sometimes employed in harvesting, as also are the Indians from the reservations, but not to any great extent.

Saddle horses may be bought for from $80 to $100 ; farm horses, from $100 to $125 ; draught horses, from $150 to $200; mules, $250 to $350 a pair; yoke cattle, $100 per pair; milch cows, $40 to $50 for good stock; sheep, $1.50 to $2.50 ; mutton sheep, $2.50 to $3.50 ; beef-cattle, per pound, six and a half to seven cents ; fat hogs, seven cents.

The price of farming land varies from five to fifty dollars per acre. Farms may be rented on very good terms. We know of one gentleman who purchased an improved farm near McMinnville, in Yamhill County, last spring, at fifteen dollars per acre, there being between three and four hundred acres in the place. He agreed to take eight hundred bushels of wheat for the rent of it; and the farmer who hired it not only paid his rent, which, at the present high price of wheat, amounted to considerably more than eight hundred dollars, but had wheat enough left to make the first payment on a farm for himself, besides supplying his family for the year. It is impossible for any but a poor farmer to be poor in pocket in a country like this.

To the above general account of Oregon farming, we add a few items about trade and revenue.

By the politeness of the Custom House officials in Astoria we are placed in receipt of the following statistics of the imports, exports, and clearances at that port from January 1st, 1871, to August 25th, 1871: Total value of exports to foreign countries, $36,167. Included are—to Peru, lumber, 391 M, $4,100; spars, etc., $388. To China, lumber, 388 M, $3,496; spars, etc., $469. To Hawaiian Islands, lumber, 229 M, $2,189; spars, etc., $438. The Knappton Mills ship full cargoes of lumber to San Francisco every month, but they do not report at the Custom-house, and we have no account of them. Total value of imports from foreign countries, $143,425. Number of vessels in foreign trade, cleared, 23; tons, 11,451. Number of vessels cleared in coast trade, 267; tons, 122,914.

The monthly report of the Chief of .the Bureau of Statistics, No. 11, shows the following imports and exports for Oregon during the eleven months ending May 31, 1871:

	Imports.	Domestic Exports.	Foreign Exports.
Merchandise	$4,708,909 79	$3,962,859 21	$129,104 70
Specie and bullion	192,798 86	750,623 11	122,925 71

Eleven months ending May 31, 1870:

	Imports.	Domestic Exports.	Foreign Exports.
Merchandise	$3,965,863 62	$3,476,813 26	$148,060 48
Specie and bullion	254,713 05	382,007 99	129,906 84

The valuation of domestic exports is given in specie.

The amount of revenue assessed during the six months ending June 30, 1871, is $57,510; of this amount, $27,765 was derived from incomes for the year. The monthly average of bank capital employed is $935,065. The monthly average of deposits in banks is $1,654,498. The total valuation of property in the Portland District, upon which taxes for school purposes will be raised, is $6,035,525. The total school tax on this is $21,124.33¾.

All business in Oregon and Washington is transacted on a gold basis. When legal tenders are used, they pass at their value in gold, as determined each day by the market quotations. A statute of the Oregon Legislature provides for the enforcement of contracts to pay in gold coin. The legal rate of interest is ten per cent. per annum; or, by express agreement, it may be made one per cent. per month, but not more.

Beyond the manufacture of lumber, flour, woolen goods, staves, linseed-oil, wagons, soap, common pottery, cabinet furniture, agricultural implements, stoves, steam-engines, and other iron works, there is nothing produced in the way of manufactures worthy of mention. The country waits for capital to bring out its almost unlimited resources in this direction, and for cheap labor to make it available.

Most kinds of mechanical labor find employment at from three to five dollars per day. Brick-layers and stone-masons get six dollars; machinists, four dollars; common laborers, one dollar and three-fourths; domestic servants, twelve to twenty-five dollars per month. All kinds of food are cheaper in Oregon and Washington than in the Atlantic States, and of a better quality. There is less adulteration in imported articles; while fruits and vegetables are both plentiful and excellent, and may be enjoyed fresh almost the whole year round.

CHAPTER XXXIV.

To such persons as may be thinking of purchasing farming lands in Oregon and Washington, we address this chapter. It is useless to look for Government land in the Wallamet Valley. All that portion of the valley which is open prairie was taken up long ago, including the School land. All the good land in the foot-hills on each side of the valley is covered by railroad and other road-grants. Therefore, to get a farm in the Wallamet Valley, the purchaser must deal with the original claimants of the level prairie, or with the owner of the road-grants. Of the first class, land may be obtained at all prices, ranging from five to fifty dollars. Of the railroad companies, land may be purchased on favorable terms, where surveyed; and settled upon by pre-emption where unsurveyed—the companies standing in the place of Government toward the settler. Only one hundred and sixty acres can be taken by pre-emption, or sold to one person. At the offices of these companies are maps, and descriptions from the surveyor's notes, of every separate parcel of land, with its valuation, which ranges in general from two dollars and a half to twelve dollars per acre.

In the valleys of Umpqua and Rogue rivers, there is more land not yet taken up; still, not a great deal, except over toward the coast. But all along the coast are large tracts of Government land, principally timbered, with occasional small prairies and creek-bottoms, which can be purchased for one dollar and a quarter per acre.

In Western Washington, along the coast, and in the northern portion, as well as at some places near the Sound, there is plenty of public land. But the greatest bodies of public land are east of the Cascades, where millions of acres of excellent soil await settlement.

The Pre-emption and Homestead laws of the United States are as follows:

"PRE-EMPTION.—Every person, being the head of a family, or widow, or single man over the age of twenty-one years, and being a citizen of the United States, or who shall have declared his intention to become a citizen, is allowed by law to make a settlement on any public land of the United States not appropriated or reserved. In the case of unsurveyed lands, legal inception by actual settlement will take place, but no proceeding toward completion of title can be had until after the land has been surveyed and the surveys returned to the District Land Office. The settler is obliged to erect a dwelling, occupy and improve the land, and make it his or her home. But no person can obtain the benefit of more than one pre-emption right, and no person who is the owner of 320 acres of land in any State or Territory, or who shall abandon his residence on his own land, to live on the public land, can acquire any right of pre-emption. Where the tract on which settlement is made has once been *offered* at public sale, a declaratory statement as to the fact of settlement must be made at the Land Office within thirty days from the date of settlement, and within one year from that date proof of residence and cultivation must be made, and the land paid for. Where the tract has been surveyed but *not* offered at public sale, the claimant must file his statement within three months from the date of settlement, and make proof and payment before the day designated by the President for the public sale of the lands.

The quantity of land allowed to one settler by pre-emption, is one quarter-section, or 160 acres, and the price to be paid, is

$1.25 per acre, except in the case of *alternate* sections embraced in any railroad reservation, which is $2.50 per acre.

Should the settler die before establishing his claim within the period limited by law, the title may be perfected by the executor, administrator, or one of the heirs, by making the requisite proof of settlement and paying for the land.

In the case of a settlement made on *unsurveyed* lands, the claimant must file notice of settlement within three months after the receipt of the township plat at the District Land Office, and make proof and payment as required in the case where surveys had been made previous to settlement.

HOMESTEADS.—The Homestead Law gives to every citizen of the United States, or foreigner declaring his intention to become such, the right to a homestead on *surveyed* lands. This is conceded to the extent of one quarter-section, or 160 acres, of land not embraced within the limits of railroad or other reservation, or eighty acres, when the location is made on alternate sections embraced *within* such reserves. To obtain homesteads the party must make affidavit that he is the head of a family, or a single man over twenty-one years of age; that he is a citizen or has declared his intention to become one, and that the location is made for his exclusive use and benefit for actual settlement and cultivation. The fees and expenses connected with the location of a homestead in Oregon are twenty-two dollars when the full amount of land is taken, or eleven dollars if half the quantity allowed by law is located. On making the affidavit before the Register and payment of the fees, a duplicate receipt will be given, which vests an inceptive right in the settler, and upon faithful observance of the law, which requires continuous settlement and cultivation for the period of five years, and upon proper proof of that fact to the Land officers within two years after the time has expired, certificates will be issued as a basis of a complete title to the land.

Where a homestead settler dies before the consummation of his claim, the heirs may continue the settlement and obtain title upon requisite proof at the proper time.

A homestead settler can not sell his claim until after his title is complete, but he can at any time relinquish his claim by surrendering his receipts, after which he is not allowed to make another settlement under the Homestead Law.

A settlement made under the Pre-emption Law may be changed to a homestead entry, if no adverse right intervenes.

If the homestead settler does not wish to remain five years on his tract, the law permits him to pay for it with cash, at the prescribed rates for claims taken by pre-emption, and upon proof of settlement and cultivation from date of entry to time of payment.

Lands obtained under the Homestead Law are exempt from liability for debts contracted prior to the issuing of a complete title by the Government.

Another method of obtaining Government lands is by 'private entry,' and applies only to such lands as have been *offered* at public sale and remain unsold. In this case payment in cash or land warrants can be made at once and a complete title obtained without delay, other than the time necessary to transmit the papers to the General Land Office and receive the patent in return. The price of land at 'private entry' is $1.25 per acre, except in the case of reserved sections: that is $2.50 per acre. At cash entry any quantity can be taken that is desired. In Eastern Oregon there is no land subject to 'private entry,' but in Western Oregon there is still a considerable amount.

There are three Land Offices in Oregon for the transaction of business connected with the disposal of Government lands: one at Oregon City, in the Wallamet Valley; one at Roseburg, in the Umpqua Valley; and one at La Grande, in Grand Ronde Valley, Eastern Oregon. The Surveyor-General's office is at Eugene City, in Lane County.

The Land Offices for Washington Territory are at Olympia, at Vancouver, and at Walla Walla."

It is only in the last four or five years that Oregon has thoroughly realized the importance of railroads to progress. But since fully awakened, great strides have been made toward connecting this remotest State of the Union with California and the East. The first railroads built on the soil of Oregon and Washington were five miles of portage around the Cascades of the Columbia, about 1853, and fifteen miles around the portage at the Dalles, in 1862, by the Oregon Steam Navigation Company. The first was a rude tramway only, until the increasing business on the river made a locomotive railway justifiable and necessary.

In 1869 the Oregon and California Railroad, from Portland to Sacramento, was commenced; but so late in the season that but twenty miles of road were completed that year. The following year it reached Albany, seventy-eight miles from Portland, and in 1871 had been pushed as far south as Eugene City, at the head of the Wallamet Valley. By the time these pages are in print, it will have been completed to Oakland, in the Umpqua Valley, one hundred and eighty-two miles from the starting-point at East Portland. At the same time, it is advancing from the south, being now completed to Tehama, one hundred and twenty-three miles north of Sacramento. Thus the six hundred miles of staging between Portland and Sacramento are being rapidly reduced, so that in another year only a day of staging will remain to vary the monotony of railroad travel.

This road receives from the General Government a grant of land amounting to 12,800 acres per mile, and becomes proprietor of nearly all the unclaimed lands in the valleys through which it passes. By very just and equitable regulations, however, the owners of these lands, called "The European and Oregon Land Company," have placed it in the power of actual settlers to select and occupy homesteads on their lands, and pay for them on exceedingly easy terms.

The second great railway in Oregon is the Oregon Central Railroad, commencing at Portland, on the west side of the Wallamet, and running to Junction City, at a point between Corvallis and Eugene. This road will also control a large amount of land; and will have a branch to the Columbia River at Astoria, or at some point in Columbia County, or both, opening up a great extent of valuable timber, mineral, and farming lands.

Twenty miles of this road are completed, and the cars were running to Cornelius, between Hillsboro and Forest Grove, in January, 1872.

The third great road, which as yet is only projected, is from the Dalles to a point on the Union Pacific, near Salt Lake City. This road, when built, will command the trade of the rich mineral districts of Southern Idaho and Eastern Oregon. Another road is talked of, from the head of the Wallamet Valley, *via* Diamond Peak Pass, or thereabout, across the Klamath country and the Humboldt Valley, to a junction with the Union Pacific.

The railroads of Washington Territory are also making good progress. The Northern Pacific, traversing the continent, from Lake Superior to Puget Sound and the Columbia River, has already completed its first twenty-five mile section on that portion of the line between the Columbia and the Sound. This road has secured a grant from the General Government of land equivalent to 25,600 acres per mile through the Territories, and 12,800 per mile through the States. If by pre-emption, settlement under the Homestead Law, or other cause, the Company are not able to obtain the quantity of land per mile which its charter entitles it to, it may make up the deficiency outside the twenty-mile limit of its land-grant.

But this Company also, like "The European and Oregon Land Company," have so systematized and facilitated the business of land sales to actual settlers, as to make it even easier for a man to select land to his liking, than it would be without the Company's assistance. The terms offered are also easy and equitable.

The second railroad of importance in Washington

Territory is that one now being built between Walla Walla and Wallula, on the Columbia River. This road will furnish an outlet for the Walla Walla Valley, adding greatly to its commercial importance and agricultural development. As the distance is only thirty miles, it is expected that this road will be completed during 1872.

These various railroads, together with the navigable waters of the Columbia and Wallamet rivers and Puget Sound, furnish, or soon will furnish, easy communication to and from almost every portion of Oregon and Washington. Only South-eastern Oregon, without navigable waters, is left to the slow locomotion of freight-wagons and stage-coaches. This, however, will not long remain so when the lines of road already commenced have drawn to themselves the population which they are sure to bring. As the circles ever widen on the water where a stone has been dropped, so the ever-widening waves of population will succeed where great railways penetrate a fertile country with a genial climate, and agreeable scenery.

Two routes are open connecting Portland with the Pacific Railroad and the East. The first *via* the Oregon and California Railroad, and the Oregon and California Stage Company's line to Sacramento. The second is *via* the Columbia River and the North-western Stage Company's line. Passengers can leave the river at Dalles, or at Umatilla, and find coaches in waiting which take them across the country to Kelton on the Central Pacific, *via* Le Grand and Baker City in Eastern Oregon, and Boise City in Idaho. To the tourist this route offers many attractions, from the peculiar scenery of the Blue Mountains and the Snake River Valley. The cost of the journey either way is about the same ; but the longer stage-ride by the latter route would cause it to be avoided by families and invalids.

From New York to San Francisco, by railroad, there are three classes of fares, ranging from $136 to $100, and $60. Sleeping-berths and meals are extra, making a first-class fare, with all the extras, cost about $180. But if passengers are provided with lunch-baskets, and dispense with sleeping-cars, and with baggage exceeding one hundred pounds, they need not spend much money over and above their fares. There is generally room enough, in the second and third-class cars, for those who are provided with a board of the proper length, to bridge the space between two seats, thus improvising, with a pair of blankets, quite a comfortable bed.

Arrived at San Francisco, the traveler has choice between the steamers of the North Pacific Transportation Company—paying a fare of $30 for first-class accommodations ; or of $15 for steerage passage, meals and baggage free ; or the overland route, by railway and stage, at $45 fare to Portland, and meals extra.

Those who prefer the steamer route from New York to San Francisco will find first-class fares ranging from $125 to $170 ; and steerage fare, $60. This includes all expenses, except such as might be occasioned by detention on the Isthmus.

Travelers and immigrants not coming to San Francisco, but bound to Montana, Idaho, and Eastern Oregon, will find stages awaiting them at Corinne, for Montana ; and at Kelton, for Idaho and Eastern Oregon. Corinne is eight hundred and fifty-seven miles east of San Francisco ; and Kelton seven hundred and ninety miles. The reduction of fare to immigrants would amount to between $20 and $30 ; and to first and second-class passengers in proportion. The stage-fare from Kelton to Umatilla, on the Columbia River, is $60, coin; time, four days; fifty pounds of baggage allowed; meals extra. Except in cases where immigrants are furnished with wagons and teams of their own, it is quite as cheap, and decidedly easier, to go to San Francisco, and thence to Portland by steamer. Persons living in the interior, away from the principal lines of travel, will find it to their advantage to send to New York, Chicago, or Omaha, for through tickets. The particular route desired to travel should be carefully specified, and the money sent through some banking-house.

From Europe to San Francisco the following is the cost of travel, estimated in gold coin:

"The Liverpool and Great Western Steamship Company sell through tickets to San Francisco, from the several European seaport towns, at rates as follows:

TO SAN FRANCISCO, FROM

	Cabin.	Steerage.
Liverpool and Queenstown	$210 00	$84 00
Hamburg, Amsterdam, Rotterdam, Harlingen, and Antwerp	224 00	90 00
Copenhagen, Gothenburg, Christiania, Bergen, Havre, Paris, Manheim	230·00	94 00

Children under twelve, half-price; under one year, $3 50.

Passengers are forwarded from New York by the boats of the Pacific Mail Steamship Company, *via* Panama. Steamers leave Liverpool and Queenstown once a week. Steerage passengers are supplied on the ocean passage with medical attendance and good, substantial food, free of cost. Owing to the fluctuations in gold in New York, the cost of forwarding passengers from that point to San Francisco is not always the same: hence the through rates from Europe are liable to some variation, though not more than a few dollars; and in any case, emigrants from Europe will find this much the cheaper route.

In case emigrants from Europe should prefer to cross the American continent by rail, the following rates of fare to the

United States by the several steamship lines will enable them to estimate the cost; railroad fare from New York and other cities heretofore given.

By the North German Lloyd Steamship line (payable in gold):

	Adult.	Children 1 to 10.	Children under 1 year.
From Bremen, Southampton, and Havre to Baltimore, cabin....	$100 00	$50 00	$2 00
Ditto, steerage..................	40 00	20 00	2 00

By the "Anchor Line" of steamers from Glasgow to New York (Steerage, payable in gold):

TO NEW YORK.

From Glasgow, Londonderry, Liverpool, and Queenstown..$34 00
  Children from 1 to 12 years, half-fare; under one year, $5.
From Hamburg, Antwerp, Rotterdam, or Havre........ 40 00
From Denmark, Norway, Sweden, or Paris............ 45 00
From Drontheim, Malmo, or Stavanger, $3 00 extra.
  Children, one to twelve years, half-fare; under one year, free.

From the figures given it is easy to compute the expense of travel and emigration, both from Europe and the Eastern States, to Portland as a central point.

From San Francisco to Puget Sound, semi-monthly steamers run *via* Victoria, British Columbia : time, four to five days; fare, $36, cabin; $20, steerage. From Portland to Puget Sound there are two routes : one by steamer every ten days, *via* the mouth of the Columbia River and Victoria; time, three days. Or from Portland to Monticello, in the Cowlitz Valley, by river steamers; fare, $1.50; and thence by stage to Olympia, eighty-five miles; fare, $10; time, two days.

The fare from Portland to Dalles, one hundred and twenty-four miles, by the Oregon Steam Navigation Company's steamers, is $5; to Umatilla, two hundred and twenty-one miles, $8; to Wallula, two hundred and forty-five miles, $10; to Walla Walla, thirty miles farther, by stage, $12; and to Lewiston, four hundred and eight miles, $20.

The distances from the mouth of the Wallamet to Jacksonville, in Southern Oregon, are as follows :

	Miles.		Miles.
Portland	12	Eugene	137
Milwaukie	18	Oakland	194
Oregon City	25	Roseburg	212
Salem	63	Canyonville	239
Albany	87	Grave Creek	265
Corvallis	97	Jacksonville	307

The fare by railroad and stage averages six cents a mile. The river steamers carry passengers for much less; but, being slower, can not compete with the railway for passenger travel.

The pleasantest months for tourists in Oregon and Washington are June and July. The rivers are then high; vegetation in its glory; the temperature delightful; and the skies blue and clear, affording views of the snowy peaks, which, after the forest fires begin, are often obscured by smoke.

SCHOOLS.

We find the provisions made by the State of Oregon for public schools to be unusually liberal. First, the State donates the sixteenth and thirty-sixth sections in each township, for the use of public schools; then, has the income of 500,000 acres donated by Congress, in 1841, for educational purposes. Seventy-two sections are reserved for a State University, and ninety thousand acres for an Agricultural College. The total amount of land reserved for school purposes is 4,475,966 acres. It is estimated that these lands will realize not less than $15,000,000. The School Fund is under the management of a Board of Commissioners, consisting of the Governor, Secretary of State, and State Treasurer, who loan it at the rate of ten per cent. per annum, secured by mortgage on real estate. The School lands are sold, on making application to the Clerk of the county in which they are situated, on the payment of one-third down, and the balance in two yearly payments, with interest, at ten per cent., on the notes, payable half yearly. They are valued at from $1.25 to $5.00 per acre, according to their quality and location. There will be other School lands coming into market when the surveys of now unsurveyed portions of the State are completed. The present School Fund amounts to $250,000, bringing in an annual interest of $25,000, to be divided among the several counties.

Besides the common schools supported out of this fund, and by taxation, there are eighteen other institutions of learning in Oregon, supported by tuition fees and endowments. Two of these are universities, several of them colleges, and the remainder academies and seminaries of various grades. St. Helen's Hall for young ladies, and the Bishop Scott Grammar-school for boys, are the two prominent seminaries in the city of Portland, after which the Portland Academy ranks next. All are well attended. St. Helen's Hall has an attendance of 179 young ladies, and the Bishop Scott Grammar-school numbers eighty-three pupils. The academy is not so flourishing as formerly, but still has a good many pupils of both sexes. Besides these, the three public-school buildings are well filled, and private schools find support.

Pacific University, at Forest Grove, with an endowment of $50,000, has in its academic and collegiate course about ninety pupils. Wallamet University, at Salem, also with $50,000 endowment, numbers a good many scholars, and has a medical department in a prosperous condition. Philomath College, near Corvallis, has an attendance of seventy scholars, and is favorably known. The Agricultural College is also established at Corvallis, and the Christian College at Monmouth.   Nearly every county in Western Oregon has one or more of these institutions of learning, which, if not yet rich enough to furnish every help to instruction which Eastern academies and colleges have, make a very fair showing for a State so thinly populated.

For a State with no more than ninety-five thousand inhabitants, it may be said, with truth, that the institutions of Oregon compare favorably with those of older and more populous ones.   But a mighty change is coming upon the whole of this North-west Coast within the next decade, which shall give to it a rank and importance that those not familiar with its advantages of climate and natural resources can very indifferently understand.   A country that has the Rocky Mountains for its eastern wall, the Pacific Sea for its western barrier; whose interior mountains are teeming with treasure; whose soil is seldom hardened by frost; down whose coasts sail no icebergs; whose wharves front those of China and Japan, and whose people are full of the intellectuality of the nineteenth century, will not pause nor hesitate on the road to wealth, learning, literature, or art.   Already Oregon has furnished the world a poet, whose mountain minstrelsy echoes from foreign shores.   From his heights, he

> Salutes his mountains—clouded Hood,
> St. Helen's in her sea of wood—
> Where sweeps the Oregon, and where
> White storms are in the feathered fir,
> And snowy sea-birds wheel and whir."

THE END.